Praise for *The Genius in All of Us*

'A deeply interesting and important book' *New York Times Book Review*

'*The Genius in All of Us* has quietly blown my mind.' Laura Miller, *Salon*

'A welcome new book ... compelling ... Shenk's thesis is that intellectual capacity is not a gift, fixed permanently in our cells. It's a process.' *Boston Globe*

'I wonder whether, finally, it's beginning to sink in among policymakers that the richness of people's lives depends on the richness of their environment, and not on the idea that some are doomed to be born thick. David Shenk's [book] should be read by anyone persisting with that myth.' *Guardian*

'Cogent and compelling ... [Shenk's book] will convince many readers that the conventional wisdom about talent is due to be overthrown. Shenk gets that revolution well under way.' *The Week*

'The thinking man's *Outliers*' *New York Magazine*

'Engrossing ... revives faith in not just practice and determination but also parenting and lifestyle.' *Booklist* (starred review)

'An incredibly well-researched meditation on the nature of human talent' Kevin Roberts, CEO Worldwide, Saatchi & Saatchi

'Outstanding' *Examiner.com*

'*The Genius in All of Us* will give new hope to those of us who have not yet written a classic sonata or played center field for the Yankees. With a flair for explaining scientific research, [Shenk] debunks outdated assumptions that genes are destiny and shows how environment and mindset are just as important.' *The Daily Beast* (a Book Pick)

'Shenk dissects and demolishes the notion that some people are "born geniuses" ... I hope that *The Genius in All of Us* is widely read and discussed among educators, and that all of us take a hard look at our own assumptions.' *Inside Higher Ed*

'Teachers, parents and anyone else who is guilty of setting low expectations for American boys should read *The Genius in All of Us*.' *Education Week*

'Empowering ... myth-busting ... entertaining.' *Kirkus Reviews*

'Startling' *Midwest Book Review* (Reviewer's Choice)

'Surprisingly compelling ... vivid and eloquently described ... equally suited to the bookshelf of a philosopher, educator, or popular science reader.' *Phenotype Journal*

'Shenk robustly disputes ... etically predetermi... ehind the "genius" of ... don cabbies).'

'Clear and exciting prose ... [this is the] one book that will change your thinking about intelligence, genetics, [and] the role of schools in creating learning.' *Cincinnati Metro News*

'Inspiring and liberating ... a powerful antidote to the genetic determinism rampant in the Age of the Genome.' Steven Johnson, author of *The Invention of Air, Ghost Map, Everything Bad is Good for You, Mind Wide Open, Emergence,* and *Interface Culture*

'A great book. David Shenk handily dispels the myth that one must be born a genius. From consistently whacking the ball out of the park to composing ethereal piano sonatas, Shenk convincingly makes the case for the potential genius that lies in all of us. While our genes may provide a nice runway, only hard work and unwavering focus can allow true genius to take flight.' Rudolph E. Tanzi PhD., Joseph P. and Rose F. Kennedy Professor of Neurology, Harvard Medical School

'David Shenk sweeps aside decades of misconceptions about genetics – and shows that by overstating the importance of genes, we've understated the potential of ourselves. *The Genius in All of Us* is a persuasive and inspiring book that will make you think anew about your own life and our shared future.' Daniel H. Pink, author of *Drive* and *A Whole New Mind*

'This book, both rigorous and accessible, is a close study of the idea of genius, an investigation of popular misconceptions about genetics, and an examination of the American virtue of self-determination. It is written with assurance, insight, clarity, and wit.' Andrew Solomon, author of *The Noonday Demon* (National Book Award Winner, 2001)

'Old fashioned beliefs, a desire to simplify and the remarkable successes of molecular biology led to an undue emphasis on the role of genes in the development of human intelligence. Environmental determinism exists too, but biology and psychology have moved well beyond these extreme positions. The importance of David Shenk's book is that he has made accessible to a wide audience the advances in the understanding of how each person develops. I congratulate him.' Sir Patrick Bateson FRS, Emeritus Professor of Ethology, Cambridge University; former Biological Secretary of the Royal Society; and co-author of *Design For A Life: How Behaviour Develops*

'David Shenk freshens and transforms a familiar subject to reveal all the interacting forces and factors that make us who we are.' Mark. S. Blumberg PhD, F. Wendell Miller Professor of Psychology, University of Iowa; editor-in-chief of *Behavioral Neuroscience;* and author of *Freaks of Nature: What Anomalies Tell Us about Development and Evolution*

'In clear, forceful language, backed up by a boatload of science, David Shenk delivers a message that should be read by every parent, educator, and policy-maker who cares about the future of our children. *The Genius in All of Us* convincingly debunks the "genes are destiny" argument when it comes to human talent, and will force you to rethink everything from IQ tests and twins studies to child-rearing practices. Shenk's book turns Baby Mozart on his head, and will give pause – a hopeful, empowering pause – to parents who wish to nurture excellence in their children.' Stephen S. Hall, author of *Wisdom: From Philosophy to Neuroscience*

The Genius in All of Us

WHY EVERYTHING YOU'VE BEEN TOLD ABOUT GENETICS, TALENT AND INTELLIGENCE IS WRONG

David Shenk

ICON BOOKS

Previously published in the UK in 2010 by Icon Books Ltd

This edition published in the UK in 2011 by
Icon Books Ltd, Omnibus Business Centre,
39–41 North Road, London N7 9DP
email: info@iconbooks.co.uk
www.iconbooks.co.uk

First published in the USA in 2010 by Doubleday, an imprint of
Random House, Inc.

Sold in the UK, Europe, South Africa and Asia
by Faber & Faber Ltd, Bloomsbury House,
74–77 Great Russell Street,
London WC1B 3DA or their agents

Distributed in the UK, Europe, South Africa and Asia
by TBS Ltd, TBS Distribution Centre, Colchester Road,
Frating Green, Colchester CO7 7DW

ISBN: 978-184831-218-0

Illustrations by Hadel Studio

Typeset in Plantin Light by Marie Doherty

Printed and bound in the UK by
Clays Ltd, St Ives plc

Compared with what we ought to be, we are only half awake. Our fires are damped, our drafts are checked. We are making use of only a small part of our physical and mental resources . . . Stating the thing broadly, the human individual lives far within his limits.

– WILLIAM JAMES

For my parents

Contents

Part Two: Cultivating Greatness

THE EVIDENCE

This book is presented in two parts of roughly equal length:

Part One, 'The Argument', states the main thesis of the book in narrative prose with very few footnotes.

Part Two, 'The Evidence', is an unconventionally lengthy section that provides the supporting material and source notes that clarify and amplify the argument.

(GxE)

(GxE)

(GxE)

The ARGUMENT

(GxE)

Individual differences in talent and intelligence are not pre-determined by genes; they develop over time. Genetic differences do play an important role, but genes do not determine complex traits on their own. Rather, genes and the environment interact with each other in a dynamic process that we can never fully control but that we can strongly influence. No two people will ever have exactly the same potential, but very few of us actually come to know our own true limits. Speaking broadly, limitations in achievement are not due to inadequate genetic assets, but to our inability, so far, to tap into what we already have.

Introduction

The Kid

Baseball legend Ted Williams was one in a million, widely considered the most 'gifted' hitter of his time. 'I remember watching one of his home runs from the bleachers of Shibe Park,' John Updike wrote in *The New Yorker* in 1960. 'It went over the first baseman's head and rose meticulously along a straight line and was still rising when it cleared the fence. The trajectory seemed qualitatively different from anything anyone else might hit.'

In the public imagination, Williams was almost a god among men, a 'superhuman' endowed with a collection of innate physical gifts, including spectacular eye-hand coordination, exquisite muscular grace, and uncanny instincts. 'Ted just had that natural ability,' said Hall of Fame second baseman Bobby Doerr. 'He was so far ahead of everybody in that era.' Among other traits, Williams was said to have laser-like eyesight, which enabled him to read the spin of a ball as it left the pitcher's fingers and to gauge exactly where it would pass over the plate. 'Ted Williams sees more of the ball than any man alive,' Ty Cobb once remarked.

But all that innate miracle-man stuff – it was all 'a lot of bull', said Williams. He insisted his great achievements were simply the sum of what he had put into the game. 'Nothing except practice, practice, practice will bring out that ability,' he explained. 'The reason I saw things was that I was so intense . . . It was [super] discipline, not super eyesight.'

Is that possible? Could a perfectly ordinary man actually train himself to be a dazzling phenomenon? We all recognise the virtues of practice and hard work, but truly, could any amount of effort transform the clunky motions of a so-so amateur sportsman into the majestic swing of Tiger Woods or the gravity-defying leap of Michael Jordan? Could an ordinary brain ever expand enough to conjure the far-flung curiosities and visions of Einstein or Matisse? Is true greatness obtainable from everyday means and everyday genes?

Conventional wisdom says no, that some people are simply born with certain gifts while others are not; that talent and high intelligence are somewhat scarce gems, scattered throughout the human gene pool; that the best we can do is to locate and polish these gems – and accept the limitations built into the rest of us.

But someone forgot to tell Ted Williams that talent will out. As a boy, he wasn't interested in watching his natural abilities unfurl passively like a flower in the sunshine. He simply wanted – needed – to be the best hitter baseball had ever seen, and he pursued that goal with appropriate ferocity. 'His whole life was hitting the ball,' recalled a boyhood friend. 'He always had that bat in his hand . . . And when he made up his mind to do something, he was going to do it or know the reason why.'

At San Diego's old North Park field, two blocks from his modest childhood home, friends recall Williams hitting baseballs every waking hour of every day, year after year after year. They describe him slugging balls until their outer shells literally wore off, swinging even splintered bats for hours upon hours with blisters on his fingers and blood dripping down his wrists. A working-class kid with no extra pocket change, he used his own lunch money to hire schoolmates to shag (chase and catch) balls so that he could keep swinging. From age six or seven, he would swing the bat at North Park field all day and night, swing until the city turned off the lights; then he'd walk home and swing a roll of newspaper in front of a mirror until he fell asleep. The next day, he'd do it all over again. Friends say he attended school only to play on the team. When baseball season ended and the other kids moved on to basketball and football, Williams stuck with baseball. When other boys started dating girls, Williams just kept hitting balls in North Park field. In order to strengthen his sight, he would walk down the street with one eye covered, and then the other. He even avoided movie theatres because he'd heard it was bad for the eyes. 'I wasn't going to let anything stop me from being the hitter I hoped to be,' Williams later recalled. 'Looking back . . . it was pretty near storybook devotion.'

In other words, he worked for it, fiercely, single-mindedly, far beyond the norm. 'He had one thought in mind and he always followed it,' said his high school coach Wos Caldwell.

Greatness was not a *thing* to Ted Williams; it was a *process.*

This didn't stop after he got drafted into professional baseball. In Williams's first season with the minor league San Diego Padres, coach Frank Shellenback noticed that his new recruit was always the first to show up for practice in the morning and the last to leave at night. And something more curious: after each game, Williams would ask the coach for the used game balls.

'What do you do with all these baseballs?' Shellenback finally asked Williams one day. 'Sell them to kids in the neighbourhood?'

'No sir,' replied Williams. 'I use them for a little extra hitting practice after supper.'

Knowing the rigours of a full practice day, Shellenback found the answer hard to swallow. Out of a mix of suspicion and curiosity, he later recalled, 'I piled into my car after supper [one night] and rode around to Williams's neighbourhood. There was a playground near his home, and sure enough, I saw The Kid himself driving those two battered baseballs all over the field. Ted was standing close to a rock which served as [home] plate. One kid was pitching to him. A half dozen others were shagging his drives. The stitching was already falling apart on the baseballs I had [just] given him.'

Even among the pros, Williams's intensity stood so far outside the norm that it was often uncomfortable to witness up close. 'He discussed the science of hitting ad nauseam with teammates and opposing players,' write biographers Jim Prime and Bill Nowlin. 'He sought out the great hitters of the game – Hornsby, Cobb and others – and grilled them about their techniques.'

He studied pitchers with the same rigour. '[After a while], pitchers figure out [batters'] weaknesses,' said Cedric Durst, who played on the Padres with Williams. 'Williams wasn't like that . . . Instead of them figuring Ted out, he figured them out. The first time Ted saw [Tony] Freitas pitch, we were sitting side by side on the bench and Ted said, "This guy won't give me a fast ball I can hit. He'll waste the fast ball and try to make me hit the curve. He'll get behind on the count, then throw me the curve." And that's exactly what happened.'

Process. After a decade of relentless effort on North Park field, and four impressive years in the minors, Williams came into the major leagues in 1939 as an explosive hitter and just kept getting better and better and better. In 1941, his third season with the Boston Red Sox, he became the only major league player in his era – and the last in the twentieth century – to bat over 0.400 for a full season.

The next year, 1942, Ted Williams enlisted in the navy as an aviator. Tests revealed his vision to be excellent, but well within ordinary human range.

. . .

Something crazy happened to the world's violinists in the twentieth century: they got better faster than their peers had in previous centuries.

We know this because we have lasting benchmarks, like the effervescent Paganini Violin Concerto no. 1 and the concluding movement of the Bach Violin Partita no. 2 in D Minor – fourteen minutes of virtually impossible violin work. Both pieces were considered nearly unplayable in the eighteenth century but are now played routinely and well by a large number of violin students.

How did this happen? And how have runners and swimmers

become so much faster, and chess and tennis players become so much more skillful? If humans were fruit flies, with a new generation appearing every eleven days, we might be tempted to chalk it up to genetics and rapid evolution. But evolution and genes don't work like that.

There is an explanation, a simple and a good one, but its implications are radical for family life and for society. It is this: some people are training harder – and smarter – than before. We're better at stuff because we've figured out how to *become* better.

Talent is not a thing; it's a *process.*

This is not at all how we're used to thinking about talent. With phrases like 'he must be gifted', 'good genes', 'innate ability', and 'natural-born [runner/shooter/talker/painter]', our culture regards talent as a scarce genetic resource, a *thing* that one either does or does not possess. IQ and other 'ability' tests codify this view, and schools build curricula around it. Journalists and even many scientists consistently validate it. This gene-gift paradigm has become a central part of our understanding of human nature. It fits with what we have been taught about DNA and evolution: *Our genes are blueprints that make us who we are. Different genes make us into different people with different abilities.* How else could the world end up with such varied individuals as Michael Jordan, Bill Clinton, Ozzy Osbourne, and you?

But the whole concept of genetic giftedness turns out to be wildly off the mark – tragically kept afloat for decades by a cascade of misunderstandings and misleading metaphors. In recent years, a mountain of scientific evidence has emerged that overwhelmingly suggests a completely different paradigm: not talent scarcity, but latent talent abundance. In this conception, human talent and intelligence are not permanently in short supply like fossil fuel, but potentially plentiful like wind power. The problem isn't our inadequate genetic assets, but our inability, so far, to tap into what we already have.

This is not to say that we don't have important genetic differences among us, yielding advantages and disadvantages. Of course we do, and those differences have profound consequences.

But the new science suggests that few of us know our true limits, that the vast majority of us have not even come close to tapping what scientists call our 'unactualised potential'. It also suggests a profound optimism for the human race. 'We have no way of knowing how much unactualised genetic potential exists,' writes Cornell University developmental psychologist Stephen Ceci. Therefore it becomes logically impossible to insist (as some have) on the existence of a genetic underclass. Most underachievers are very likely not prisoners of their own DNA, but rather have so far been unable to tap into their true potential.

This new paradigm does not herald a simple shift from 'nature' to 'nurture'. Instead, it reveals how bankrupt the phrase 'nature versus nurture' really is and demands a whole new consideration of how each of us becomes us. This book begins, therefore, with a surprising new explanation of how genes work, followed by a detailed look at the newly visible building blocks of talent and intelligence. Taken together, a new picture emerges of a fascinating developmental process that we can influence – though never fully control – as individuals, as families, and as a talent-promoting society. While essentially hopeful, the new paradigm also raises unsettling new moral questions with which we all will have to grapple.

It would be folly to suggest that anyone can literally do or be anything, and such is not this book's intent. But the new science tells us that it's equally foolish to think that mediocrity is built into most of us, or that any of us can know our true limits before we've applied enormous resources and invested vast amounts of time. Our abilities are not set in genetic stone. They are soft and sculptable, far into adulthood. With humility, with hope, and with extraordinary determination, greatness is something to which any kid – of any age – can aspire.

PART ONE

THE MYTH OF GIFTS

CHAPTER ONE

Genes 2.0

How Genes Really Work

Contrary to what we've been taught, genes do not determine physical and character traits on their own. Rather, they interact with the environment in a dynamic, ongoing process that produces and continually refines an individual.

The sun begins to rise over an old river town, and through a fifth-floor window of University Hospital, a newborn cries out her own birth announcement. Her new, already sleep-deprived parents hold her tightly and simply stare, partly in disbelief that this has actually happened, partly in awe of what lies ahead. As she develops, who will she look like? What will she be like? What will be her strengths, her weaknesses? Will she change the world or just scrape by? Will she run a quick mile, paint a new idea, charm her friends, sing for millions? Will she have any talent for anything?

Only the years will tell. For right now, the parents don't really need to know the final outcome – they just need to know what sort of difference they can make. How much of their newborn daughter's personality and abilities are already predetermined? What portion is still up for grabs? What ingredients can they add, and what tactics should they avoid?

The fuzzy mix of hope, expectation, and burden begins . . .

TONY SOPRANO: And to think [I'm] the cause of it.
DR. MELFI: How are you the cause of it?
TONY SOPRANO: It's in his blood, this miserable fucking existence. My rotten fucking putrid genes have infected my kid's soul. That's my gift to my son.

Genes can be scary stuff if you don't understand them. In 1994, psychologist Richard Herrnstein and policy analyst Charles Murray warned in their bestselling book *The Bell Curve* that we live in an increasingly stratified world where the 'cognitive elite' – those with the best genes – are more and more isolated from the cognitive/genetic underclass. 'Genetic partitioning', they called it. There was no mistaking their message:

> The irony is that as America equalises the [environmental] circumstances of people's lives, the remaining differences in intelligence are increasingly determined by differences in genes . . . Putting it all together, success and failure in the American economy, and all that goes with it, are increasingly a matter of the genes that people inherit.

Stark and terrifying – and thankfully quite mistaken. The authors had fundamentally misinterpreted a number of studies, becoming convinced that roughly 60 per cent of each person's intelligence comes directly from his or her genes. But genes don't work that way. 'There are no genetic factors that can be studied independently of the environment,' explains McGill University's Michael Meaney, one of the world's leading experts on genes and development. 'And there are no environmental factors that function independently of the genome. [A trait] emerges only from the *interaction* of gene and environment.'

While Herrnstein and Murray adhered to a particular ideological agenda, they also seem to have been genuinely hobbled in their analysis by a common misunderstanding of how genes work. We've all been taught that we inherit complex traits like intelligence straight from our parents' DNA in the same way we inherit simple traits like eye colour. This belief is continually reinforced by the popular media. As an illustration, *USA Today* recently explained heredity in this way:

> Think of your own genetic makeup as the hand of cards you were dealt at conception. With each conception in a family comes a new shuffling of the deck and a new hand. That's partly why little Bobby sleeps through the night as a baby, always behaves and seems to love math, while brother Billy is colicky, never listens and already is the head of a gang in kindergarten.

Genes dictate. Genes instruct. Genes determine. For more than a century, this has been the widely accepted explanation of how each of us becomes us. In his famous pea-plant experiments of the 1850s and '60s, Gregor Mendel demonstrated that basic traits like seed shape and flower colour were reliably passed from one generation to the next through dominant and recessive 'heritable factors' (Mendel's phrase before the word 'gene' was introduced). After eight years and twenty-eight thousand plants, Mendel had proved the existence of genes – and seemed to prove that genes alone determined the essence of who we are. Such was the unequivocal interpretation of early-twentieth-century geneticists.

That notion is with us still. 'Genes set the stage,' affirms *USA Today*. The environment has an impact on all of our lives, to be sure, but genes come first; they set specific lower and upper limits of each person's potential abilities. *Where did your brother get that amazing singing voice? How did you get so tall? Why can't I dance? How is she so quick with numbers?*

'It's in the genes,' we say.

That's what *The Bell Curve* authors thought, too. None of these writers realised that over the last two decades Mendel's ideas have been thoroughly upgraded – so much so that one large group of scientists now suggests that we need to wipe the slate clean and construct an entirely new understanding of genes.

This new vanguard is a loose-knit group of geneticists, neuroscientists, cognitive psychologists, and others, some of whom call themselves developmental systems theorists. I call them *interactionists* because of their emphasis on the dynamic interaction between genes and the environment. Not all of the interactionists' views have yet been fully accepted, and they freely acknowledge their ongoing struggle to articulate the full implications of their findings. But it already seems very clear that these implications are far-reaching and paradigm-shifting.

To understand interactionism, you must first try to forget everything you think you know about heredity. 'The popular conception of the gene as a simple causal agent is not valid,' declare geneticists Eva Jablonka and Marion Lamb. 'The gene cannot be seen as an autonomous unit – as a particular stretch of DNA which always produces the same effect. Whether or not a length of DNA produces anything, what it produces, and where and when it produces it may depend on other DNA sequences and on the environment.'

Though Mendel couldn't detect it with his perfectly calibrated pea-plant hybrids, genes are not like robot actors who always say the same lines in the exact same way. It turns out that they interact with their surroundings and can say different things depending on whom they are talking to.

This obliterates the long-standing metaphor of genes as blueprints with elaborate predesigned instructions for eye colour, thumb size, mathematical quickness, musical sensitivity, etc. Now we can come up with a more accurate metaphor. Rather than finished blueprints, genes – all twenty-two thousand of them[1] – are more like volume knobs and switches. Think of a giant control board inside every cell of your body.

Many of those knobs and switches can be turned up/down/on/off at any time – by another gene or by any minuscule environmental input. This flipping and turning takes place constantly. It begins the moment a child is conceived and doesn't

[1] Estimates of the actual number of genes vary.

stop until she takes her last breath. Rather than giving us hardwired instructions on how a trait must be expressed, this process of gene-environment interaction drives a unique developmental path for every unique individual.

The new interactionists call it 'GxE' for short. It has become central to the understanding of all genetics. Recognition of GxE means that we now realise that genes powerfully influence the formation of all traits, from eye colour to intelligence, but rarely dictate precisely what those traits will be. From the moment of conception, genes constantly respond to, and interact with, a wide range of internal and external stimuli – nutrition, hormones, sensory input, physical and intellectual activity, and other genes – to produce a unique, custom-tailored human machine for each person's unique circumstance. Genes matter, and genetic differences will result in trait differences, but in the final analysis, each of us is a dynamic system, a creature of development.

This new dynamic model of GxE (genes multiplied by environment) is very different from the old static model of G+E (genes plus environment). Under the old paradigm, genes came first and set the stage. They dealt each of us our first hand of cards, and only afterward could we add in environmental influences.

The new model begins with interaction. There is no genetic foundation that gets laid before the environment enters in; rather, genes express themselves strictly in accordance with their environment. Everything that we are, from the first moment of conception, is a result of this process. We do not *inherit* traits directly from our genes. Instead, we *develop* traits through the dynamic process of gene-environment interaction. In the GxE world, genetic differences still matter enormously. But, on their own, they don't determine who we are.

In fact, you did not even inherit your blue eyes or brown hair from your parents' genes. Not directly.

This may sound crazy at first, because of how thoroughly we've been indoctrinated with Mendelian genetics. The reality turns out to be much more complicated – even for pea plants. Many scientists have understood this much more complicated truth for years but have had trouble explaining it to the general public. It is indeed a lot harder to explain than simple genetic determinism.

. . .

To understand genes more fully, we first need to take a step back and explain what they actually do:

Genes direct the production of proteins.

Each of our cells contains a complete double strand of DNA, which in turn contains thousands of individual genes. Each gene initiates the process of assembling amino acids into proteins. Proteins are large, specialised molecules that help create cells, transport vital elements, and produce necessary chemical reactions.

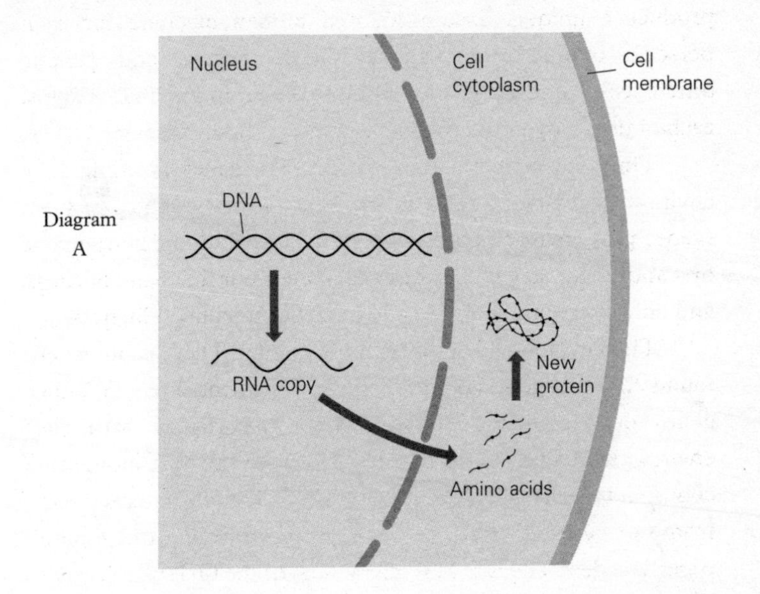

There are many different protein types, and they provide the building blocks of everything from muscle fibre to eyeball collagen to haemoglobin. We are, each one of us, the sum of our proteins.

Genes contain the instructions for the formation of those proteins, and they direct the protein-building process (Diagram A).

But . . . genes are *not* the only things influencing protein construction. It turns out that the genetic instructions themselves are influenced by other inputs. Genes are constantly activated and deactivated by environmental stimuli, nutrition, hormones, nerve impulses, and other genes (Diagram B).

This explains how every brain cell and hair cell and heart cell in your body can contain *all* of your DNA but still perform very specialised functions. It also explains how a tiny bit of genetic diversity goes a very long way: human beings are distinct from one another not just because of our relatively few genetic differences, but also because every moment of our ongoing lives actively influences our own genetic expression.

Think of GxE as baking a cake, suggests Cambridge

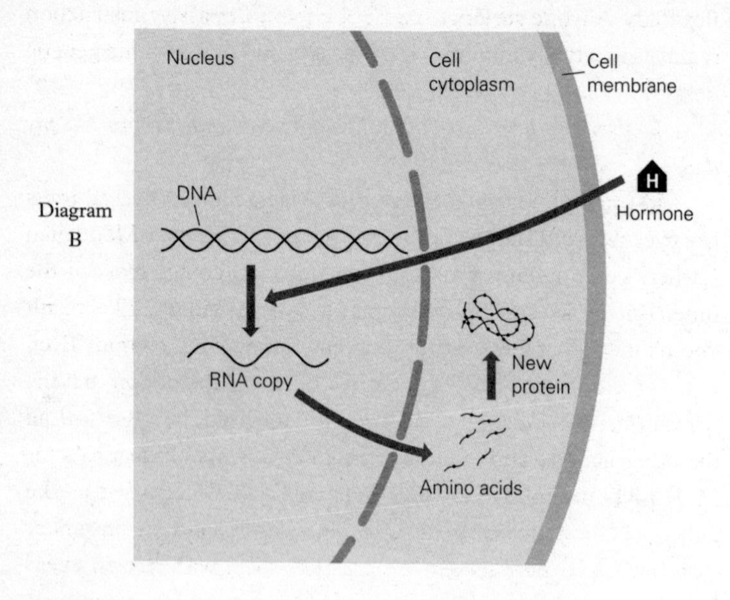

University biologist Patrick Bateson. A hundred cooks may start out with nearly the same ingredients but will in the end produce very different cakes. While the slight difference in ingredients guarantees that differences will exist, it doesn't dictate what those differences will be. The actual end-result differences arise out of the process. 'Development is chemistry,' says Bateson, 'and the end product cannot simply be reduced to its ingredients.'

Similarly, the mere presence of a certain gene does not automatically produce a specific type or number of proteins. First, every gene has to be activated – switched on, or 'expressed' – in order to initiate protein construction.

Further, geneticists have recently discovered that some genes – we don't yet know how many – are versatile. In some cases, the exact same gene can produce different proteins depending on how and when it is activated.

All of this means that, on their own, most genes cannot be counted on to directly produce specific traits. They are active participants in the developmental process and are built for flexibility. Anyone seeking to describe them as passive instruction manuals is actually minimising the beauty and power of the genetic design.

So why do I have brown eyes like my mum and red hair like my dad?

In practical terms, there are many elementary physical traits like eye, hair, and skin colour where the process is near Mendelian – where certain genes produce predictable outcomes most of the time. But looks can be deceiving; a simple Mendel-like result doesn't mean that there wasn't gene-environment interaction. 'Even in the case of eye colour,' says Patrick Bateson, 'the notion that the relevant gene is *the* [only] cause is misconceived, because [of] all the other genetic and environmental ingredients.' Indeed, Victor McKusick, the Johns Hopkins geneticist widely regarded as the father of clinical medical genetics, reminds us that in some instances 'two blue-eyed parents can produce children with brown eyes.' Recessive genes cannot explain such an event; gene-environment interaction can.

When it comes to more complex traits like physical coordination, personality, and verbal intelligence, gene-environment interaction inevitably moves the process even further away from simple Mendelian patterns.

What about single genetic mutations that predictably cause diseases such as Huntington's disease?

Single-gene diseases do exist and account for roughly 5 per cent of the total disease burden in developed countries. But it's important not to let such diseases give the wrong impression about how healthy genes work. 'A disconnected wire can cause a car to break down,' explains Patrick Bateson. 'But this does not mean that the wire by itself is responsible for making the car move.' Similarly, a genetic defect causing a series of problems does not mean that the healthy version of that gene is single-handedly responsible for normal function.

Helping the public understand gene-environment interaction is a particular burden, because it is so enormously complex. It will never have the same easy, snap-your-fingers resonance that our old (misleading) understanding of genes had for us. Given that, the interactionists are lucky to have Patrick Bateson on their side. A former biological secretary to the Royal Society of London and one of the world's leading public educators about heredity, Bateson also carries a powerful symbolic message with his surname. It was his grandfather's famous cousin, William Bateson, who, a century ago, first coined the word 'genetics' and helped popularise the earlier, simpler notion of genes as self-contained information packets that directly produce traits. Now the third-generation Bateson is helping to significantly update that public understanding.

'Genes store information coding for the amino acid sequences of proteins,' explains Bateson. 'That is all. They do not code for parts of the nervous system and they certainly do not code for particular behaviour patterns.'

His point is that genes are several steps removed from the process of trait formation. If someone is shot dead with a Smith & Wesson handgun, no one would accuse the guy running the blast furnace that transformed the iron ore into pig iron – which was subsequently transformed into steel and later poured into various molds before being assembled into a Smith & Wesson handgun – of murder. Similarly, no gene has explicit authorship of good or bad vision, long or short legs, or affable or difficult personality. Rather, genes play a crucial role throughout the process. Their information is translated by other actors in the cell and influenced by a wide variety of other signals coming from outside the cell. Certain types of proteins are then formed, which become other cells and tissues and ultimately make us who we are. The step-by-step distance between a gene and a trait will depend on the complexity of the trait. The more complex the trait, the farther any one gene is from direct instruction. This process continues throughout one's entire life.

Height can provide a terrific insight into the gene-environment dynamic. Most of us think of height as being more or less directly genetically determined. The reality is so much more interesting. One of the most striking early hints of the new understanding of development as a dynamic process emerged in 1957 when Stanford

School of Medicine researcher William Walter Greulich measured the heights of Japanese children raised in California and compared them to the heights of Japanese children raised in Japan during the same time period. The California-raised kids, with significantly better nourishment and medical care, grew an astonishing five inches taller on average. Same gene pool, different environment – radically different stature. Greulich didn't realise this at the time, but it was a perfect illustration of how genes really work: not dictating any predetermined forms or figures, but interacting vigorously with the outside world to produce an improvised, unique result.

It turns out that a wide variety of environmental elements will affect the genetic expression of height: a single case of diarrhoea or measles, for example, or deficiencies in any one of dozens of nutrients. In Western cultures of the twenty-first century, we tend to assume a natural evolutionary trend of increased height with each generation, but in truth human height has fluctuated dramatically over time in specific response to changes in diet, climate, and disease. Most surprising of all, height experts have determined that, biologically, very few ethnic groups are truly destined to be taller or smaller than other groups. While this general rule has some exceptions, 'by and large,' sums up *The New Yorker*'s Burkhard Bilger, 'any population can grow as tall as any other . . . Mexicans ought to be tall and slender. Yet they're so often stunted by poor diet and diseases that we assume they were born to be small.'

Born to be small. Born to be smart. Born to play music. Born to play basketball. It's a seductive assumption, one that we've all made. But when one looks behind the genetic curtain, it most often turns out not to be true.

Another stunning example of the gene-environment interactive dynamic arrived, coincidentally, just one year after Greulich's Japanese height study. In the winter of 1958, Rod Cooper and John Zubek, two young research psychologists at the University of Manitoba, devised what they thought was a classic nature/nurture experiment about rat intelligence. They started with newborn rat pups from two distinct genetic strains: 'Maze-bright' rats, which had consistently tested well in mazes over many generations, and 'Maze-dull' rats, which had consistently tested poorly in those same mazes, making an average of 40 per cent more mistakes.

Then they raised each of these two genetic strains in three very different living conditions:

> **Enriched environment:** featuring walls painted in rich, bright patterns and many stimulating toys: ramps, mirrors, swings, slides, bells, etc.
> **Normal environment:** with ordinary walls and a moderate amount of exercise and sensory toys.
> **Restricted environment:** essentially rat slums with nothing but a food box and a water pan; no toys or anything else to stimulate their bodies or minds.

In broad terms, it seemed easy enough to predict the outcome: each strain of rat would get a little smarter when raised in the enriched environment and get a little dumber when raised in the poor environment. They expected to have a graph that looked something like this:

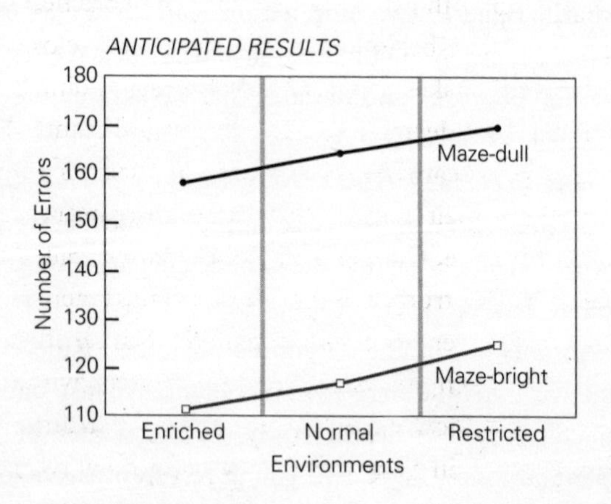

Instead, the results looked like this:

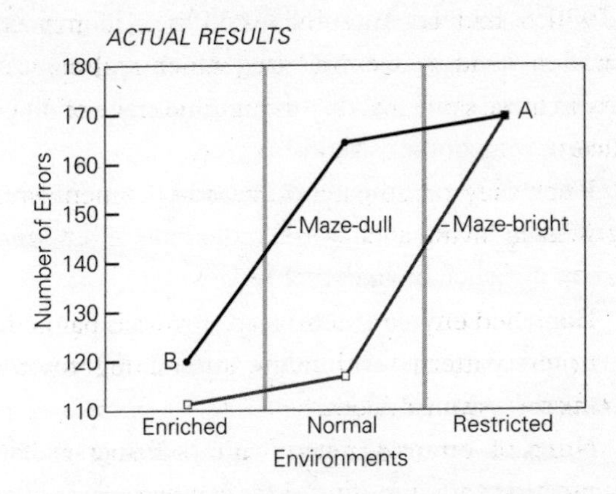

The final data were quite shocking. Under normal conditions, the Maze-bright rats had consistently outperformed the Maze-dull rats. But in both extreme environments, they performed virtually the same. The Maze-bright rats raised in the restricted environment made almost exactly the same number of mistakes as the Maze-dull rats raised in the restricted environment (point A, above). In other words, when raised in an impoverished environment, all the rats seemed equally dumb. Their 'genetic' differences disappeared.

The same thing happened with the enriched environment. Here, the Maze-bright rats also made very close to the same number of mistakes as the Maze-dull rats (point B, above – the difference was deemed statistically insignificant). Raised in an exciting, provocative environment, all the rats seemed equally smart. Again, their 'genetic' differences disappeared.

At the time, Cooper and Zubek didn't really know what to make of it. The truth was that these original 'genetic' differences hadn't really ever been purely genetic. Rather, they had been a function of each strain's GxE development within its original environment. Now, when developing within different environments, each strain was producing very different results. And in the case of both the enriched and restricted environments, the different genetic strains turned out to be a lot more alike than they had previously seemed.

In the decades that followed, the Cooper-Zubek study emerged as 'a classic example of gene-environment interaction,' according to Penn State developmental geneticist Gerald McClearn. Many other scientists agree.

Over this same time period, hundreds of examples emerged that gradually forced a wholesale rethinking of how genes operate. Almost in disbelief, biologists observed that

- the temperature surrounding turtle and crocodile eggs determined their gender
- young, yellow-skinned grasshoppers became permanently black skinned for camouflage if exposed to a blackened (burnt) environment at a certain age
- locusts living in a crowded environment developed vastly more musculature (suitable for migration) than locusts living in less crowded conditions

In these and so many other instances, environment A seemed to produce one kind of creature while environment B produced another creature entirely. This level of trait modification was simply impossible to comprehend under the old G+E idea that genes directly determined traits. The new facts demanded a whole new explanation of how genes function.

In 1972, Harvard biologist Richard Lewontin supplied a critical clarification that helped his colleagues understand GxE. The old nature-and-nurture view featured a one-way, additive sequence like this:

Genes→proteins→cells→traits
↗
Environment

Genes trigger the production of proteins, which guide the functions of cells, which, with some input from the outside world, form traits.

The new GxE was a much more dynamic process, with every input at every level influencing every other input:

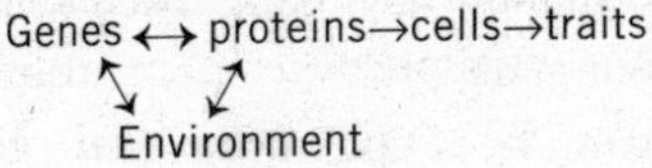

Genes ↔ proteins→cells→traits
↖↘ ↗↙
Environment

Genes, proteins, and environmental signals (including human behaviour and emotion) constantly interact with one another, and this interactive process influences the production of proteins, which then guide the functions of cells, which form traits.

Note the influence-arrows moving in both directions in the second sequence. 'Biologists have come to realise that if one changes *either* the genes *or* the environment, the resulting behaviour can be dramatically different,' explains City University of New York evolutionary ecologist Massimo Pigliucci. 'The trick, then, is not in partitioning causes between nature and nurture, but in [examining] the way genes and environments interact dialectically to generate an organism's appearance and behaviour.'

The great irony, then, of our endless efforts to distinguish nature from nurture is that we instead need to do exactly the opposite: to try to understand precisely how nature and nurture *interact*. Precisely which genes do get switched on, and when, and how often, and in what order, will make all the difference in the function of each cell – and the traits of the organism.

'In each case,' explains Patrick Bateson, 'the individual animal starts its life with the capacity to develop in a number of distinctly different ways. Like a jukebox, the individual has the potential to play a number of different developmental tunes. But during the course of its life it plays only one tune. The particular developmental tune it does play is selected by [the environment] in which the individual is growing up.'

From that first moment of conception, then, our temperament, intelligence, and talent are subject to the developmental process. Genes do not, on their own, make us smart, dumb, sassy, polite, depressed, joyful, musical, tone-deaf, athletic, clumsy, literary, or incurious. Those characteristics come from a complex interplay within a dynamic system. Every day in every way you are helping to shape which genes become active. Your life is interacting with your genes.

The dynamic model of GxE turns out to play a critical role in everything – your mood, your character, your health, your lifestyle, your social and work life. It's how we think, what we eat, whom we marry, how we sleep. The catchy phrase 'nature/nurture' sounded good a century ago, but it makes no sense today, since there are no truly separate effects. Genes and the environment are as inseparable and inextricable as letters in a word or parts in a car. We cannot embrace or even understand the new world of talent and intelligence without first integrating this idea into our language and thinking.

We need to replace 'nature/nurture' with 'dynamic development'.

How did Tiger Woods end up with the most dependable stroke and the toughest competitive drive in the history of golf? Dynamic development. How did Leonardo da Vinci develop into an unparalleled artist, engineer, inventor, anatomist, and botanist? Dynamic development. How did Richard Feynman advance from a boy with a merely good IQ score to one of the most important thinkers of the twentieth century? Dynamic development.

Dynamic development is the new paradigm for talent, lifestyle, and well-being. It is how genes influence everything but strictly determine very little. It forces us to rethink everything about ourselves, where we come from, and where we can go. It promises that while we'll never have true control over our lives, we do have the power to impact them enormously. Dynamic development is why human biology is a jukebox with many potential tunes – not specific built-in instructions for a certain kind of life, but built-in capacity for a variety of possible lives. No one is genetically doomed to mediocrity.

Dynamic development was one of the big ideas of the twentieth century, and remains so. Once our brand-new parents in University Hospital understand its implications for their newborn girl, it will affect how they live, how they parent, and even how they vote.

CHAPTER TWO

Intelligence Is a Process, Not a Thing

Intelligence is not an innate aptitude, hardwired at conception or in the womb, but a collection of developing skills driven by the interaction between genes and environment. No one is born with a predetermined amount of intelligence. Intelligence (and IQ scores) can be improved. Few adults come close to their true intellectual potential.

[Some] assert that an individual's intelligence is a fixed quantity which cannot be increased. We must protest and react against this brutal pessimism.

– ALFRED BINET,
inventor of the original IQ test, 1909

London is a taxi driver's nightmare, a preposterously large and convoluted urban jungle built up chaotically over some fifteen hundred years. This is not a city built neatly on a grid, like Manhattan or Barcelona, but a crude patchwork of ancient Roman, Viking, Saxon, Norman, Danish, and English settlement roads, all laid on top of and around one another. Within a six-mile radius of Charing Cross Station, some twenty-five thousand streets connect and bisect at every possible angle, dead-ending into parks, monuments, shops, and private homes. In order to be properly licensed, London taxi drivers must learn *all* of these driving nooks and crannies – an encyclopaedic awareness known proudly in the trade as 'The Knowledge'.

The good news is that, once learned, The Knowledge becomes literally embedded in the taxi driver's brain. That's what British neurologist Eleanor Maguire discovered in 1999 when she and her colleagues conducted MRI scans on London cabbies and compared them with the brain scans of others. In contrast with noncabbies, experienced taxi drivers had a greatly enlarged posterior hippocampus – that part of the brain that specialises in recalling spatial representations. On its own, that finding proved nothing; theoretically, people born with larger posterior hippocampi could have innately better spatial skills and therefore be more likely to become cabbies. What made Maguire's study so striking is that she then correlated the size of the posterior hippocampi directly with each driver's experience: the longer the driving career, the larger the posterior hippocampus. That strongly suggested that spatial tasks were actively changing cabbies' brains. 'These data,' concluded Maguire dramatically, 'suggest that the changes in hippocampal grey matter . . . are acquired.'

Further, her conclusion was perfectly consistent with what others have discovered in recent studies of violinists, Braille readers, meditation practitioners, and recovering stroke victims: that specific parts of the brain adapt and organise themselves in response to specific experience. 'The cortex has a remarkable capacity for remodelling after environmental change,' reported Harvard psychiatrist Leon Eisenberg in a comprehensive review.

This is our famous 'plasticity': every human brain's built-in capacity to become, over time, what we demand of it. Plasticity does not mean that we're all born with the exact same potential. Of course we are not. But it does guarantee that no ability is fixed. And as it turns out, plasticity makes it virtually impossible to determine any individual's true intellectual limitations, at any age.

. . .

How smart can you become? What are you capable of intellectually? For many decades, psychologists thought they had a reliable instrument to answer this question: the Stanford-Binet

Intelligence Scales, otherwise known as the IQ test. This combination of tests, measuring language and memory skills, visual-spatial abilities, fine motor coordination, and perceptual skills, was said by its inventor, Lewis Terman, to reveal a person's 'original endowment' – his innate intelligence.

> Psychological methods of measuring intelligence [have] furnished conclusive proof that native differences in endowment are a universal phenomenon.
>
> – LEWIS TERMAN, *Genetic Studies of Genius*, 1925

A prominent research psychologist at Stanford University, Terman was part of a well-established movement convinced that intelligence was an inborn asset, inherited through genes, fixed at birth, and stable throughout life. Revealing each person's intelligence would, they believed, help individuals find their rightful places in society and help society run more efficiently. The movement's original founder had been Francis Galton, a half cousin to, and peer of, Charles Darwin in mid-nineteenth-century England. After Darwin published *On the Origin of Species* in 1859, Galton immediately sought to further define natural selection by arguing that differences in human intellect were strictly a matter of biological heredity – what he called the 'hereditary transmission of physical gifts'.

Galton did not share the cautious scientific temperament of his cousin Darwin but was a forceful advocate for what he believed in his gut to be true. In 1869, he published *Hereditary Genius*, arguing that smart, successful people were simply 'gifted' with a superior biology. In 1874, he introduced the phrase 'nature and nurture' (as a rhetorical device to favour nature). In 1883, he invented 'eugenics', his plan to maximise the breeding of biologically superior humans and minimise the breeding of biologically inferior humans. All of this was in service to his conviction that natural selection was driven exclusively by biological heredity and that the environment was just a passive bystander. In fact, it was actually Galton, not Darwin, who laid the conceptual groundwork for genetic determinism.

A few decades later, though, Galton's followers ran into a serious problem: they couldn't actually locate the natural, innate intelligence they were arguing for. In fact, they couldn't even agree how to define it. Was intelligence a facility in logical reasoning? Spatial visualisation? Mathematical abstraction? Physical coordination? 'In truth,' lamented British psychologist and statistician Charles Spearman, '[the word] "intelligence" has become a mere vocal sound, a word with so many meanings that finally it has none.'

In 1904, Spearman introduced his own solution to this problem: there must be a single 'general intelligence' (*g* for short), he theorised, a centralised entity of intellectual skills. And though it couldn't be measured directly – and still can't – Spearman argued that *g* could be detected statistically, through a correlation of different measures. Using his 'simple' mathematical formula

$$\left(r_{pq} = \frac{r_{pq} - r_{pv} \bullet r_{qv}}{\sqrt{(1 - r^2_{pv})(1 - r^2_{qv})}} \right)$$

he established a correlation between school marks, teachers' subjective assessments, and peers' assessments of 'common sense'. This correlation, Spearman argued, proved the existence of a central, inborn thinking ability. '*G* is, in the normal course of events, determined innately,' Spearman declared. 'A person can no more be trained to have it in higher degree than he can be trained to be taller.'

In 1916, Stanford's Lewis Terman produced a practical equivalent of *g* with his Stanford-Binet Intelligence Scales (adapted from an earlier version by French psychologist Alfred Binet) and declared it to be the ideal tool to determine a person's native intelligence. While some immediately saw through Terman's claim,[2]

[2] 'Without offering any data on all that occurs between conception and the age of kindergarten,' *New Republic* editor Walter Lippmann wrote in 1922, '[Terman and colleagues] announce . . . that they are measuring the hereditary mental endowment of human beings. Obviously, this is not a conclusion obtained by research. It is a conclusion planted by the will to believe.'

most greeted IQ with enthusiasm. The U.S. Army quickly adopted a version for recruiting, and schools followed. Everything about IQ's crispness and neat classifications fit perfectly with the American hunger for enhanced social, academic, and business efficiencies.

Unfortunately, that same meritocracy movement carried an underbelly of profound racism in which alleged proof of biological superiority of white Protestants was used to keep blacks, Jews, Catholics, and other groups out of the higher ranks of business, academia, and government. In the early 1920s, the National Intelligence Test (a precursor to the SAT) was designed by Edward Lee Thorndike, an ardent eugenicist determined to convince college administrators how wasteful and socially counterproductive it would be to provide higher education to the masses. 'The world will get better treatment,' Thorndike declared, 'by trusting its fortunes to its 95 or 99-percentile intelligences.' Interestingly, just a few years later, the SAT's creator, Princeton psychologist Carl Brigham, disavowed his own creation, writing that all intelligence tests were based on 'one of the most glorious fallacies in the history of science, namely that the tests measured native intelligence purely and simply without regard to training or schooling.'

Aside from overt ethnic discrimination, the real and lasting tragedy of IQ and other intelligence tests was the message they sent to every individual – including the students who scored well. That message was: *your intelligence is something you were given, not something you've earned.* Terman's IQ test easily tapped into our primal fear that most of us are born with some sort of internal restraining bolt allowing us to think only so deeply or quickly. This is extraordinary, considering that, at its core, IQ was merely a population-sorting tool.

IQ scores do not actually report how well you have objectively mastered test material. They merely indicate how well you have mastered it compared to everyone else. Given that it simply ranked individuals in a population, it is particularly sad to look back and see that Lewis Terman and colleagues actually recommended that individuals identified as 'feebleminded' by his test be removed from

Distribution of IQ Scores

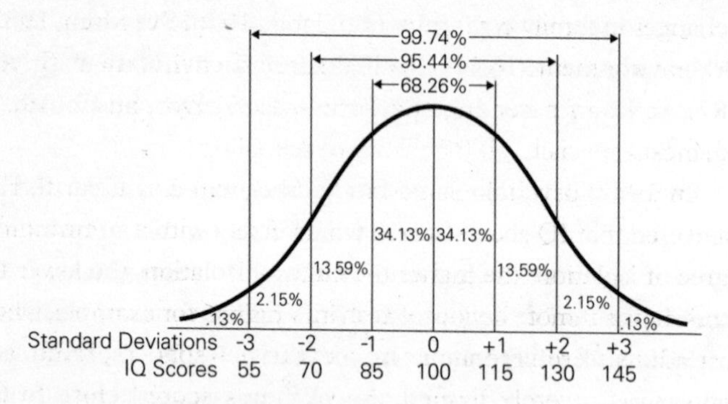

IQ scores rank academic achievement within each age group. The scoring is weighted so that 100 always marks the precise centre of the population curve, indicating that precisely one half of the age group has performed better and the other half has performed worse. A score of 115 indicates that almost 16 per cent performed better. A score of 70 indicates that almost 98 per cent performed better, and so on.

society and that anyone scoring less than 100 be automatically disqualified from any prestigious position. To automatically dismiss the worth of anyone scoring below 100 was to mistake relative value for absolute value. It was like saying that, out of any one hundred oranges, fifty are never going to taste very good.

IQ did succeed admirably in one regard: it standardised academic comparisons and thus became a very useful way of comparing academic achievement across schools, states, even nations. Any school principal, governor, etc., would certainly want to know whether his students were underperforming or outperforming the national average. Further, these tests measured achievement broadly enough to predict generally how test takers would fare in the future, compared to others.

But measuring achievement was enormously different from pinpointing individual capacity. Predicting how most kids will do is entirely different from declaring what any particular kid *can* do. 'Stability,' Exeter University's Michael Howe points out, 'does not imply unchangeability.' And indeed, individual IQ scores are quite alterable if a person gets the right push. 'IQ scores,' explains Cornell

University's Stephen Ceci, 'can change quite dramatically as a result of changes in family environment (Clarke, 1976; Svendsen, 1982), work environment (Kohn, 1981), historical environment (Flynn, 1987), styles of parenting (Baumrind, 1967; Dornbusch, 1987), and, most especially, shifts in level of schooling.'

In 1932, psychologists Mandel Sherman and Cora B. Key discovered that IQ scores correlated inversely with a community's degree of isolation: the higher the cultural isolation, the lower the scores. In the remote hollow of Colvin, Virginia, for example, where most adults were illiterate and access to newspapers, radio, and schools was severely limited, six-year-olds scored close to the national average in IQ. But as the Colvin kids got older, their IQ scores drifted lower and lower – falling further and further behind the national average due to inadequate schooling and acculturation. (The very same phenomenon was discovered among the so-called canal boat children in Britain and in other isolated cultural pockets.) Their unavoidable conclusion was that 'children develop only as the environment demands development.'

Children develop only as the environment demands development. In 1981, New Zealand–based psychologist James Flynn discovered just how profoundly true that statement is. Comparing raw IQ scores over nearly a century, Flynn saw that they kept going up: every few years, the new batch of IQ test takers seemed to be smarter than the old batch. Twelve-year-olds in the 1980s performed better than twelve-year-olds in the 1970s, who performed better than twelve-year-olds in the 1960s, and so on. This trend wasn't limited to a certain region or culture, and the differences were not trivial. On average, IQ test takers improved over their predecessors by three points every ten years – a staggering difference of eighteen points over two generations.

The differences were so extreme, they were hard to wrap one's head around. Using a late-twentieth-century average score of 100, the comparative score for the year 1900 was calculated to be about 60 – leading to the truly absurd conclusion, acknowledged Flynn, 'that a majority of our ancestors were mentally retarded.' The so-called Flynn effect raised eyebrows throughout the world of cognitive research. Obviously, the human race had not evolved into a markedly smarter species in less than one hundred years. Something else was going on.

For Flynn, the pivotal clue came in his discovery that the increases were not uniform across all areas but were concentrated in certain subtests. Contemporary kids did not do any better than their ancestors when it came to general knowledge or mathematics. But in the area of abstract reasoning, reported Flynn, there were 'huge and embarrassing' improvements. The further back in time he looked, the less test takers seemed comfortable with hypotheticals and intuitive problem solving. Why? Because a century ago, in a less complicated world, there was very little familiarity with what we now consider basic abstract concepts. '[The intelligence of] our ancestors in 1900 was anchored in everyday reality,' explains Flynn. 'We differ from them in that we can use abstractions and logic and the hypothetical . . . Since 1950, we have become more ingenious in going beyond previously learned rules to solve problems on the spot.'

Examples of abstract notions that simply didn't exist in the minds of our nineteenth-century ancestors include the theory of natural selection (formulated in 1864), and the concepts of control group (1875) and random sample (1877). A century ago, the scientific method itself was foreign to most Americans. The general public had simply not yet been conditioned to think abstractly.

The catalyst for the dramatic IQ improvements, in other words, was not some mysterious genetic mutation or magical nutritional supplement but what Flynn described as 'the [cultural] transition from pre-scientific to post-scientific operational thinking.' Over the course of the twentieth century, basic principles of science slowly filtered into public consciousness, transforming the world we live in. That transition, says Flynn, 'represents nothing less than a liberation of the human mind.'

> The scientific world-view, with its vocabulary, taxonomies, and detachment of logic and the hypothetical from concrete referents, has begun to

> permeate the minds of post-industrial people. This has paved the way for mass education on the university level and the emergence of an intellectual cadre without whom our present civilization would be inconceivable.

Perhaps the most striking of Flynn's observations is this: 98 per cent of IQ test takers today score better than the average test taker in 1900. The implications of this realisation are extraordinary. It means that in just one century, improvements in our social discourse and our schools have dramatically raised the measurable intelligence of almost *everyone*.

So much for the idea of fixed intelligence. We know now that, even though most people's relative intellectual ranking tends to remain the same as they grow older:

- It's not biology that establishes an individual's rank to begin with (social, academic, and economic factors are well-documented contributors).
- No individual is truly stuck in her original ranking.
- Every human being (even a whole society) can grow smarter if the environment demands it.

None of this has dissuaded proponents of innate intelligence, who continue to insist that IQ's stability proves a natural, biological order of minds: the gifted few naturally ascend to greatness while those stuck at the other end of the spectrum serve as an unwanted drag on modern society. 'Our ability to improve the academic accomplishment of students in the lower half of the distribution of intelligence is severely limited,' Charles Murray wrote in a 2007 op-ed in the *Wall Street Journal*. 'It is a matter of ceilings . . . We can hope to raise [the grade of a boy with an IQ slightly below 100]. But teaching him more vocabulary words or drilling him on the parts of speech will not open up new vistas for him. It is not within his power to follow an exposition written beyond a limited level of complexity . . . [He is] not smart enough.'

'Even the best schools under the best conditions cannot repeal the limits on achievement set by limits on intelligence,' Murray says bluntly.

But an avalanche of ongoing scholarship paints a radically different, more fluid, and more hopeful portrait of intelligence.

. . .

In the mid-1980s, Kansas psychologists Betty Hart and Todd Risley realised that something was very wrong with Head Start, America's programme for children of the working poor. It manages to keep some low-income kids out of poverty and ultimately away from crime. But for a programme that intervenes at a very young age and is reasonably well run and generously funded – $7 billion annually – it doesn't do much to raise kids' academic success. Studies show only 'small to moderate' positive impacts on three- and four-year-old children in the areas of literacy and vocabulary, and no impact at all on maths skills.

The problem, Hart and Risley realised, wasn't so much with the mechanics of the programme; it was the *timing*. Head Start wasn't getting hold of kids early enough. Somehow, poor kids were getting stuck in an intellectual rut long before they got to the programme – before they turned three and four years old. Hart and Risley set out to learn why and how. They wanted to know what was tripping up kids' development at such an early age. Were they stuck with inferior genes, lousy environments, or something else?

They devised a novel (and exhaustive) methodology: for more than three years, they sampled the actual number of words spoken to young children from forty-two families at three different socioeconomic levels: (1) welfare homes, (2) working-class homes, and (3) professionals' homes. Then they tallied them up.

The differences were astounding. Children in professionals' homes were exposed to an average of more than fifteen hundred more spoken words per hour than children in welfare homes. Over one year, that amounted to a difference of nearly 8 million words, which, by age four, amounted to a total gap of 32 million words. They also found a substantial gap in tone and in the complexity of words being used.

As they crunched the numbers, they discovered a direct correlation between the intensity of these early verbal experiences and later achievement. 'We were astonished at the differences the data revealed,' Hart and Risley wrote in their book *Meaningful Differences*. 'The most impressive aspects [are] how different individual families and children are and how much and how important is children's cumulative experience before age 3.'

Not surprisingly, the psychological community responded with a mixture of interest and deep caution. In 1995, an American Psychological Association task force wrote that 'such correlations may be mediated by genetic as well as (or instead of) environmental factors.' Note 'instead of'. In 1995, it was still possible for leading research psychologists to imagine that better-off kids could be simply inheriting smarter genes from smarter parents, that spoken words could be merely a genetic effect and not a cause of anything.

Now we know better. We know that genetic factors do not operate 'instead of' environmental factors, they interact with them: GxE. Genetic differences do exist. But those differences aren't straitjackets holding us in place; they are bungee cords waiting to be stretched and stretched. When positive environmental triggers such as parental speaking are discovered, the appropriate response is not to caution against their possible irrelevance, but to embrace their influence on our genes – and our lives.

And now we know what some of those triggers are:

- **Speaking to children early and often.** This trigger was revealed in Hart and Risley's incontrovertible study and reinforced by the University of North Carolina's Abecedarian Project, which provided environmental enrichment to children from birth, with the study subjects showing substantial gains compared with a control group.
- **Reading early and often.** In 2003, a national study reported the positive influence of early parent-to-child reading, regardless of parental education level. In 2006, a similar study again found the same thing about

reading, this time ruling out any effects of race, ethnicity, class, gender, birth order, early education, maternal education, maternal verbal ability, and maternal warmth.

- **Nurturance and encouragement.** Hart and Risley also found that, in the first four years after birth, the average child from a professional family receives 560,000 more instances of encouraging feedback than discouraging feedback; a working-class child receives merely 100,000 more encouragements than discouragements; a welfare child receives 125,000 more discouragements than encouragements.
- **Setting high expectations.** As Sherman and Key found in 1932, 'children develop only as the environment *demands* development.'
- **Embracing failure.** Coaches, CEOs, teachers, parents, and psychologists all now recognise the importance of pushing their charges to the limit, and just beyond. Setbacks must be seen as learning tools rather than signs of permanent built-in limitation.
- **Encouraging a 'growth mindset'.** Stanford psychologist Carol Dweck has built her prestigious career on the importance of individuals believing that their own abilities are malleable – not fixed from birth. Many studies show that the more a person believes that abilities can be developed, the greater the success that person will eventually enjoy. (More on Dweck in chapter 5.)

Recognising the value of these and other environmental inputs doesn't take away from the importance of genetics. In the new GxE paradigm, to embrace environmental influences is also to embrace the importance of genes: Reading *expresses* genes. Speaking *expresses* genes. Mentoring *expresses* genes.

With GxE, intelligence is not a thing, but a process. Why do some kids do better in school right from the start? Why are they earlier talkers, earlier achievers, and ultimately more creatively and financially successful in their adult lives? It's because from day one, they are trained to be.

. . .

Around the same time that James Flynn was discovering his Flynn effect, and Hart and Risley were uncovering their early spoken-word effect, City University of New York research psychologist Sylvia Scribner came upon a very different (but no less striking) phenomenon that we might call 'carton calculus'. This oddity was quietly unfolding in a Baltimore dairy plant, where uneducated carton packers revealed remarkable mathematical abilities in their work. Though they were easily the least educated people in the factory, they could, without hesitation or discussion, determine exactly which of many orders to fill in precisely which sequence so as to minimise bending over and walking. For example:

> If an order called for 6 pints of whole milk, 12 pints of two-per cent milk, and 3 pints each of skim milk and buttermilk, an experienced assembler might select a case for 24 pints that was already half-filled with two-per cent milk and one-third filled with whole milk, rather than try to prepare the order from scratch with an empty case. Using the half-filled case would enable the assembler to fill the order by removing 2 pints of whole milk and adding 3 pints each of skim milk and buttermilk, for a total of only three [back] bends.
>
> Moreover, when the orders were not evenly divisible into cases, the assemblers were able to shift between different representations of the order, *a feat equivalent to shifting between different-base systems of numbers*.

The maths and mental effort involved was staggering, and yet the low-paid assemblers did this routinely all day long. 'Assemblers calculated these least-physical effort solutions even when the "saving" in moves amounted to only one unit (in orders that might total 500 units),' explained Scribner.

No signs of this ability showed up on IQ scores, maths tests, or school grades. By any conventional academic measure, these labourers were thoroughly unintelligent. And yet, when the highly educated white-collar workers from the same factory occasionally filled in with assembler tasks, they couldn't begin to match the case-filling expertise of an experienced low-IQ assembler.

Halfway around the world, in Kisumu, Kenya, Yale psychologist Robert Sternberg stumbled on exactly the same phenomenon in 2001 when studying the intelligence of Dholuo schoolchildren. First he measured their knowledge of local herbal remedies, then tested them according to their Western curriculum. Surprisingly, Sternberg found a 'significantly negative' correlation. 'The better the children did on the indigenous tacit knowledge,' he noted, 'the worse they did on the test of vocabulary used in school, and vice-versa.'

Why – and which test represented true intelligence?

Actually, none of these studies will likely come as a real shock to the reader. We're all familiar with the notion of 'street smarts' as opposed to 'school smarts'. But the Baltimore carton packers and the Kisumu schoolkids did pose a serious challenge to research psychologists adhering to traditional definitions of intelligence. As Robert Sternberg watched studies like these pile up – documenting the unusual, sometimes even untestable intelligence traits of Yup'ik Eskimo children, !Kung San hunters of the Kalahari Desert, Brazilian street youth, American horse handicappers, and Californian grocery shoppers – he realised that the lack of correlation between their expertise and IQ scores demanded nothing less than a whole new definition of intelligence.

He saw another problem, too, that reinforced this conclusion: the increasingly flimsy distinction between 'intelligence' tests and so-called achievement tests like the SAT II. The more Sternberg compared the two, the harder it was for him to find any real difference between them. Both test types measure achievements, Sternberg concluded – skills that a person has developed.

All of this finally led Sternberg – one of the leading authorities in the study of human intellect – to tear down the wall that prevented the public from understanding the truth about intelligence.

'Intelligence,' he declared profoundly in 2005, 'represents a set of competencies in development.'

In other words, intelligence isn't fixed. Intelligence isn't general. Intelligence is not a thing. Intelligence is a dynamic, diffuse, and ongoing process. This finding fits perfectly with the earlier work of Mihály Csikszentmihályi and colleagues, who concluded that 'high academic achievers are not necessarily born "smarter" than others, but work harder and develop more self-discipline.'

We can trick ourselves into thinking that measuring a person's intelligence is like measuring the length of a table. But in truth, it's more like measuring a five-year-old's weight. Whatever measurement you get applies only for today. How will that child measure up tomorrow? In large part, that is up to the child, and to all of us.

CHAPTER THREE

The End of 'Giftedness'

(and the True Source of Talent)

Like intelligence, talents are not innate gifts, but the result of a slow, invisible accretion of skills developed from the moment of conception. Everyone is born with differences, and some with unique advantages for certain tasks. But no one is genetically designed into greatness and few are biologically restricted from attaining it.

In 1980, the young Swedish psychologist Anders Ericsson found himself working with the great William Chase, one of the pioneers of cognitive psychology. Chase, at Carnegie Mellon University in Pittsburgh, helped explore the implications of chunking, the memory technique used by all human beings to convert a scattered collection of details into a single distinct memory. Phone numbers, for example, are not stored in our brains as ten separate numbers but in three easy chunks: 513-673-8754. Remembering ten unrelated items in the right order is next to impossible; remembering three is no problem. The same notion applies to remembering words, music, chess positions, or any other constellation of symbols. Great minds don't recall more raw data than others; rather, they recognise patterns faster and form chunks more efficiently.

Chunking had offered a major breakthrough in understanding how the mind works. Now Ericsson and Chase were interested in learning even more about the severe limits of short-term memory and how to circumvent them. While our long-term memory capacity is apparently limitless, new memories are almost pathetically fragile: the average healthy adult can reliably juxtapose only three or four new, unrelated items. Such a limit, noted Ericsson and Chase, 'places severe constraints on the human ability to process information and solve problems.'

But what about apparent exceptions to this rule – the handful of famous memory experts ('mnemonists') who've been able to recall prodigious amounts of new and disconnected information? Ericsson and Chase wanted to know if these remarkable performers had innate memory gifts or if they had somehow acquired their extraordinary skills. To answer that question, they embarked on an unusual and ambitious experiment.

They attempted to create a mnemonist from scratch.

Could an ordinary person's short-term memory be trained, like a juggler, to handle a much larger amount of information? There was only one way to find out. Ericsson and Chase recruited an undistinguished college student for an epic experiment. The student, known by his initials, S.F., tested normal for intelligence and normal for short-term memory performance. Memory-wise, he was just like you or me. Then they began the training. It was gruelling work. In one-hour sessions, three to five sessions per week, researchers read sequences of random numbers to S.F. at the rate of one digit per second: *2 . . . 5 . . . 3 . . . 5 . . . 4 . . . 9 . . .* At intervals, they stopped and asked him to echo their list back. 'If the sequence was reported correctly,' the researchers noted, 'the next sequence was increased by one digit; otherwise it was decreased by one digit.' *2 . . . 5 . . . 3 . . . 5 . . . 4 . . . 9 . . . 7 . . .* At the end of every session, S.F. was asked to recall as many of that day's numbers as possible. *2 . . . 5 . . . 3 . . . 5 . . . 4 . . . 9 . . . 7 . . . 6 . . .*

Instead of jumping off a bridge or transferring to another college, S.F. kept returning to the memory lab. In fact, he continued to participate most days of the week for more than two years – more than 250 hours of lab time. Why? Perhaps because he was seeing results. Almost immediately, his short-term memory performance started to improve: from seven digits to ten after a handful of

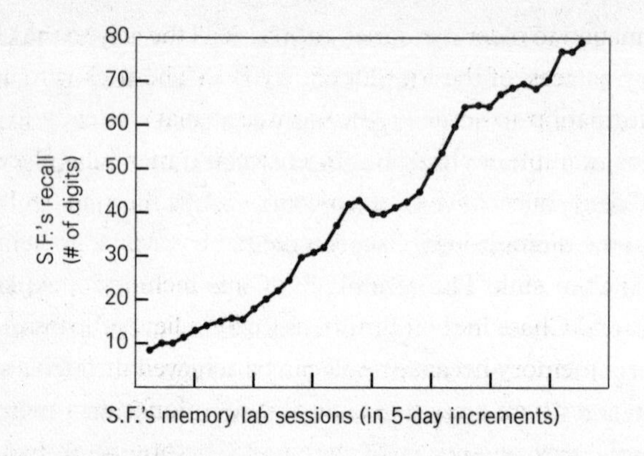

sessions, then to an amazing twenty digits after several more dozen training hours. Already he had clearly escaped the normal bounds of short-term memory. From there, the improvements continued unabated: to thirty digits, forty, fifty, sixty, seventy, and finally to a staggering eighty-plus digits before the team concluded the experiment.

S.F.'s progress is represented on the graph above.

There was no indication as the sessions ended that he had reached any sort of boundary. 'With practice,' Ericsson and Chase concluded, 'there is seemingly no limit to memory performance.'

How did he do it? Through interviews with S.F., Ericsson and Chase realised that their subject had neither tapped into a hidden memory gift nor somehow transformed the brain circuitry of his short-term memory. Rather, he had simply employed clever strategies that enabled him to get around his – and all of our – natural limits.

Here's how:

S.F. happened to be a competitive runner. Early on, after trying in vain simply to remember as many random numbers as possible, he realised that when he pictured an unconnected string of three or four digits as one single race time – for example, converting the numbers 5–2–3–4 into five minutes and twenty-three point four seconds – the numbers would come back to him quite easily.

This was not a new technique; attaching disconnected pieces

of information to older memories goes back all the way to the Greek 'memory palaces' of the fourth century B.C. The trick is to assign new information to some system or image that's already in your head. For example, a classroom teacher could mentally 'place' the face and name of each new student in a different room in her home: Lucas in the dining room; Oscar in the pantry; Malcolm standing by the kitchen sink. The advantage of this technique, explained Ericsson and Chase in their report, 'is that it relieves the burden on short-term memory because recall can be achieved through a single association with an already-existing code in long-term memory.' S.F., like every impressive mnemonist before him, had not transformed his natural memory limit; instead he had changed the way he formed new memories to take advantage of a different, less restrictive memory system.

But how did the researchers know for sure that S.F. had not actually altered his short-term memory capacity? Simple: between number sessions, they also tested him with random alphabet letters: *U . . . Q . . . B . . . Y . . . D . . . X . . .* Whenever they did this, his memory performance immediately reverted to normal. Without special mnemonic tricks and lots of contextual practice, his short-term memory was again as ordinary as yours or mine.

Ericsson and Chase published their results in the prestigious journal *Science*, and their results were subsequently corroborated many times over. They concluded:

> These data suggest that . . . it is not possible to increase the capacity of short-term memory with extended practice. Rather, increases in memory span are due to the use of mnemonic associations in long-term memory. With an appropriate mnemonic system and retrieval structure, there is seemingly no limit to improvement in memory skill with practice.

It was a double lesson: when it comes to memory skills, there is no escaping basic human biology – nor any need to. Remembering extraordinary amounts of new information simply requires the right strategies and the right amount of intensive practice, tools theoretically available to any functioning human being.

So began Anders Ericsson's remarkable talent odyssey. He quickly suspected that the importance of his discovery went far beyond mind puzzles like geometry and chess. There were implications here, he imagined, for playing the cello, shooting a basketball, painting a canvas, brewing sake, reading a CT scan – any skill where real-time performance is dependent on one's knowledge and experience. Though he couldn't be sure at the time, Ericsson suspected he had just discovered the hidden key to the veiled domains of talent and genius.

He was right.

. . .

Truly great accomplishments are inherently mysterious, awe inspiring, even intimidating to witness. What daunting thoughts course through the mind of any listener as ten-year-old Midori plays Paganini's Sauret cadenza with such startling grace and finesse? Beyond the feeling of amazement is the inevitable comparison with oneself – the acknowledgement that if you drew that same bow across those same strings on that exact same violin, such squeaks and squawks would fill the room as to make people run for cover.

By the same token, one watches David Beckham bend that ball into the goal, or Michael Jordan fly through the air towards the hoop, or Tiger Woods knock a tiny ball 325 yards to within inches of the hole, and one experiences an exhilarating but also deflating feeling: *these extraordinary performers cannot possibly belong to the same species as you or me.*

Call it the greatness gap – that sensation of an infinite and permanent chasm between ultra-achievers and mere mortals like us. Such feelings beg for a reassuring explanation: This person has something I do not have. They were born with something I wasn't born with. They are gifted.

It is an assumption built right into our culture. 'Talent' is defined in the *Oxford English Dictionary* as 'mental endowment;

natural ability' and is sourced all the way back to the parable of the talents in the book of Matthew. The words 'gifted' and 'giftedness' date back to the seventeenth century. The term 'genius', as it is currently defined, goes back to the tail end of the eighteenth century.

Recent centuries are peppered with evocative statements reinforcing the idea of inborn gifts:

- 'Poets and musicians are born,' declared the poet Christian Friedrich Schubart in 1785.
- 'Musical genius is that inborn, inexplicable gift of Nature,' insisted the composer Peter Lichtenthal in 1826.
- 'Don't ask, young artist, "what is genius?"' proclaimed Jean-Jacques Rousseau in 1768. 'Either you have it – then you feel it yourself, or you don't – then you will never know it.'

In the twentieth century, the presumed source of a person's natural endowment shifted from God-given to gene-given, but the basic notion of giftedness remained substantially the same. Exceptional abilities were things bestowed upon a very lucky person.

Notably, Friedrich Nietzsche dissented along the way. In his 1878 book *Menschliches, Allzumenschliches* (*Human, All-Too-Human*), he described greatness as being steeped in a process, and of great artists being tireless participants in that process:

> Artists have a vested interest in our believing in the flash of revelation, the so-called inspiration . . . [shining] down from heavens as a ray of grace. In reality, the imagination of the good artist or thinker produces continuously good, mediocre, and bad things, but his judgment, trained and sharpened to a fine point, rejects, selects, connects . . . All great artists and thinkers [are] great workers, indefatigable not only in inventing, but also in rejecting, sifting, transforming, ordering.

As a vivid illustration, Nietzsche cited Beethoven's sketchbooks, which reveal the composer's slow, painstaking process of testing and tinkering with melody fragments like a chemist constantly pouring different concoctions into an assortment of beakers.

Beethoven would sometimes run through as many as sixty or seventy different drafts of a phrase before settling on the final one. 'I make many changes, and reject and try again, until I am satisfied,' the composer once remarked to a friend. 'Only then do I begin the working-out in breadth, length, height and depth in my head.'

Alas, neither Nietzsche's nuanced articulation nor Beethoven's candid admission caught on with the general public. Instead, the simpler and more alluring idea of giftedness prevailed and has since been carelessly and breathlessly reinforced by biologists, psychologists, educators, and the media. Three essential ingredients have kept it alive:

1. **The unexplained phenomena of child prodigies and 'savants':** tiny Mozarts and Midoris in possession of spectacular abilities that seem to come from nowhere.
2. **The myth of genes as blueprints:** a simple and compelling account of where giftedness comes from, not substantially refuted until recently.
3. **No compelling alternative:** no sweeping contrary evidence from scientists, and no effective rhetorical substitutes from writers.

All of which left 'giftedness' as the only acceptable explanation for exceptional ability. Few psychologists or educators resisted the temptation to use it as a shorthand when discussing talents.

But Anders Ericsson did resist.

After his 1980 memory experiments, the old giftedness dogma just didn't seem to make sense anymore. Though he was not a geneticist and, at the time, had no way to know just how bankrupt the gene-blueprint myth truly was, he defied convention and proposed a radical new conception of talent: talent is not the *cause* but the *result* of something. It doesn't create a process but is the end result of that process. If true, this would mean that high achievement in many physical and creative realms is much more attainable among human beings than is implied by the notion of giftedness.

Over the following three decades, Ericsson and colleagues invigorated the largely dormant field of expertise studies in order to test this idea, examining high achievement from every possible angle: memory, cognition, practice, persistence, muscle response, mentorship, innovation, attitude, response to failure, and on and on. They studied golfers, nurses, typists, gymnasts, violinists, chess players, basketball players, and computer programmers.

They also examined many of the vivid historical myths of talent and genius, poking through the clichés to see if any clear-eyed lessons could be drawn. Standing above all other giftedness legends, of course, was that of the mystifying boy genius Wolfgang Amadeus Mozart, alleged to be an instant master performer at age three and a brilliant composer at age five. His breathtaking musical gifts were said to have sprouted from nowhere, and his own father promoted him as the 'miracle which God let be born in Salzburg'.

The reality about Mozart turns out to be far more interesting and far less mysterious. His early achievements – while very impressive, to be sure – actually make good sense considering his extraordinary upbringing. And his later undeniable genius turns out to be a wonderful advertisement for the power of process.

Mozart was bathed in music from well before his birth, and his childhood was quite unlike any other. His father, Leopold Mozart, was an intensely ambitious Austrian musician, composer, and teacher who had gained wide acclaim with the publication of the instruction book *Versuch einer grüendlichen Violinschule* (*A Treatise on the Fundamental Principles ofViolin Playing*). For a while, Leopold had dreamed of being a great composer himself. But on becoming a father, he began to shift his ambitions away from his own unsatisfying career and onto his children – perhaps, in part, because his career had already hit a ceiling: he was vice-kapellmeister (assistant music director); the top spot would be unavailable for the foreseeable future.

Uniquely situated, and desperate to make some sort of lasting mark on music, Leopold began his family musical enterprise even before Wolfgang's birth, focusing first on his daughter Nannerl. Leopold's elaborate teaching method derived in part from the Italian instructor Giuseppe Tartini and included highly nuanced techniques:

> [Leopold] advocated the so-called 'Geminiani grip' for greater left hand facility and good intonation and . . . recommended that each finger be left in place until required to move – a procedure which would also have contributed to a more effective legato . . . he placed emphasis on the freedom of the right elbow and hand, stressing the need to keep the bowing arm low but recommending that the violin be tilted towards the E-string side – thereby allowing for a freer wrist action.

As a court composer, Leopold Mozart was an ordinary creature of his place and era. As a music teacher, though, he was centuries ahead of his time. Eventually, his focus on technique and his impulse to teach very young children would be widely adopted by Shinichi Suzuki and other twentieth-century instructors. But this was quite rare in the eighteenth century; only a handful of families in the world could have conceivably enjoyed the same level of in-family attention, expertise, and ambition. With first-rate home instruction and exceptional amounts of practice, Nannerl Mozart became, over the course of a few years, a dazzling pianist and violinist – *for her age*. (As a rule, child prodigies are not adult-level innovators but masters of technical skill; their spellbinding quality comes out of natural comparison with other children's skills, not because they truly compare to the best adult performers in their field.)

Then came Wolfgang. Four and a half years younger than his sister, the tiny boy got everything Nannerl got – only much earlier and even more intensively. Literally from his infancy, he was the classic younger sibling soaking up his big sister's singular passion.

As soon as he was able, he sat beside her at the harpsichord and mimicked notes that she played. Wolfgang's first pings and plucks were just that. But with a fast-developing ear, deep curiosity, and a tidal wave of family know-how, he was able to click into an accelerated process of development.

As Wolfgang became fascinated with playing music, his father became fascinated with his toddler son's fascination – and was soon instructing him with an intensity that far eclipsed his efforts with Nannerl. Not only did Leopold openly give preferred attention to Wolfgang over his daughter; he also made a career-altering decision to more or less shrug off his official duties in order to build an even more promising career for his son. This was not a quixotic adventure. Leopold's calculated decision made reasonable financial sense in two ways: First, Wolfgang's youth made him a potentially lucrative attraction. Second, as a male, Wolfgang had a promising, open-ended future musical career. As a woman in eighteenth-century Europe, Nannerl was severely limited in this regard.

From the age of three, then, Wolfgang had an entire family driving him to excel with a powerful blend of instruction, encouragement, and constant practice. He was expected to be the pride and financial engine of the family, and he did not disappoint. In his performances from London to Mannheim between the ages of six and eight, he drew good receipts and high praise from noble patrons. He could play rehearsed minuets or sight-read études he had never seen before, could play the clavier with a thick cloth covering his hands and the keys, could improvise a coherent piece from a suggested theme.

Still, like his sister, the young Mozart was never a truly great adult-level instrumentalist. He was highly advanced for his age, but not compared with skillful adult performers. The tiny Mozart dazzled royalty and was at the time unusual for his early abilities. But today many young children exposed to Suzuki and other rigorous musical programmes play as well as the young Mozart did – and some play even better. Inside the world of these intensive, child-centred programmes, such achievements are now straightforwardly regarded by parents and teachers for what they are: the combined consequence of early exposure, exceptional instruction, constant practice, family nurturance, and a child's intense will to learn. Like a brilliant soufflé, all of these ingredients must be present in just the right quantity and mixed with just the right timing and flair. Almost anything can go wrong. The process is far from predictable and never in anyone's complete control.

It is a blessing for any person, at any age, to be able to bring grace and beauty into other people's lives. But such feats among children tend to cloud the judgment of adult observers, leading to what neuroscientist and musicologist Daniel J. Levitin calls 'the circular logic of talent'. 'When we say that someone is talented,' he says, 'we think we mean that they have some innate predisposition to excel, but in the end, we only apply the term retrospectively, after they have made significant achievements.'

Levitin is exactly right. A profound ambiguity swirls around the word, which perpetually confuses the issue for anyone using it. 'Talent' can be used to describe your daughter's strong interest in an activity, or what you regard as her undeveloped promise, or her developing skill, or her unexplained advantage over peers. In a culture where linguistic precision is paramount, where we have at least twenty-five different words for 'delicious' and thirteen for 'ridicule', such ambiguity is the best possible indicator of a real gap in our understanding of this powerful force in our lives. Aside from love, talent may be the most important intangible in all of human society. It is a linguistic apparition.

But what if the intangible could be made tangible? Over the last three decades, Anders Ericsson's research army has aimed to do just that. Like all good scientists, their approach has been to break down athletic, intellectual, and artistic achievements into tiny, measurable components in order to determine what separated the mediocre from the good, the good from the very good, the very good from the extraordinary. They've interviewed, taped, tabulated, and scanned. They've measured eye movements, muscle response, breaths, swings, strokes, torque, ventricular function, white matter, grey matter, and memory. They've watched people hone skills, or not, over many years' time. Over time, a picture has emerged – not nearly complete, but vivid enough to begin to see a process, to actually witness the tiny moving parts driving individual improvement. For those on their way to greatness, several themes consistently come to light:

1. **Practice changes your body.** Researchers have recorded a constellation of physical changes (occurring in direct response to practice) in the muscles, nerves, hearts, lungs, and brains of those showing profound increases in skill level in any domain.
2. **Skills are specific.** Individuals becoming great at one particular skill do not serendipitously become great at other skills. Chess champions can remember hundreds of intricate chess positions in sequence but can have a perfectly ordinary memory for everything else. Physical and intellectual changes are ultraspecific responses to particular skill requirements.
3. **The brain drives the brawn.** Even among athletes, changes in the brain are arguably the most profound, with a vast increase in precise task knowledge, a shift from conscious analysis to intuitive thinking (saving time and energy), and elaborate self-monitoring mechanisms that allow for constant adjustments in real time.
4. **Practice style is crucial.** Ordinary practice, where your current skill level is simply being reinforced, is not enough to get better. It takes a special kind of practice to force your mind and body into the kind of change necessary to improve.
5. **Short-term intensity cannot replace long-term commitment.** Many crucial changes take place over long periods of time. Physiologically, it's impossible to become great overnight.

Across the board, these last two variables – practice style and practice time – emerged as universal and critical. From Scrabble players to dart players to soccer players to violin players, it was observed that the uppermost achievers not only spent significantly more time in solitary study and drills, but also exhibited a consistent (and persistent) style of preparation that Ericsson came to call 'deliberate practice'. First introduced in a 1993 *Psychological Review* article, the notion of deliberate practice went far beyond the simple idea of hard work. It conveyed a method of continual skill improvement. 'Deliberate practice is a very special form of activity that differs from mere experience and mindless drill,' explains Ericsson. 'Unlike playful engagement with peers, deliberate practice is not inherently enjoyable. It . . . does not involve a mere execution or repetition of already attained skills but repeated attempts *to reach beyond one's current level* which is associated with *frequent failures.* Aspiring performers therefore concentrate on improving specific aspects by engaging in practice activities designed to change and refine particular mediating mechanisms, requiring problem solving and successive refinement with feedback.'

In other words, it is practice that doesn't take no for an answer; practice that perseveres; the type of practice where the individual keeps raising the bar of what he or she considers success.

How does deliberate practice actually improve one's skills? In a nutshell, our muscles and brain regions adapt to the demands that we make of them. 'Frequent intense engagement in certain types of practice activities,' writes Ericsson, 'is shown to induce physiological strain which causes biochemical changes that stimulate growth and transformation of cells, which in turn leads to associated improved adaptations of physiological systems and the brain.'

Recall Eleanor Maguire's 1999 brain scans of London cabbies, which revealed greatly enlarged representation in the brain region that controls spatial awareness. The same holds for any specific task being honed; the relevant brain regions adapt accordingly.

For deliberate practice to work, the demands have to be serious and sustained. Simply playing lots of chess or soccer or golf isn't enough. Simply taking lessons from a wonderful teacher is not enough. Simply wanting it badly enough is not enough. Deliberate practice requires a mind-set of never, ever, being satisfied with your current ability. It requires a constant self-critique, a pathological restlessness, a passion to aim consistently just beyond one's capability so that daily disappointment and failure is actually desired, and a never-ending resolve to dust oneself off and try again and again and again.

It also requires enormous, life-altering amounts of time – a daily grinding commitment to becoming better. In the long term, the results can be highly satisfying. But in the short term, from day to day and month to month, there's nothing particularly fun about the process or the substantial sacrifices involved. In studies, Ericsson found a clear distinction between leisure players, who tend to enjoy themselves casually much of the time, and dedicated achievers, who become glued to the gritty process of getting better:

> Whereas the amateur singers experienced the lesson as self-actualisation and an enjoyable release of tension, the professional singers increased their concentration and focused on improving their performance during the lesson. In their research on chess expertise, Charness et al. (1996, 2005) found that the amount of solitary chess study was the best predictor of performance during chess tournaments . . . Similar findings of the unique effectiveness of deliberate solitary practice have been reported by Duffy et al., 2004, for dart throwing. A recent study by Ward et al. (2004) demonstrated that elite level youth soccer players spent less time in playful activities than less-skilled control participants, and accrued more time spent engaged in deliberate practice.

What about those who practise regularly and strenuously, pursuing their pursuits seriously, but who do not improve significantly? Are they just missing that magic genetic spark? Not as far as Ericsson and his team can tell. 'A careful review of the published evidence on the heritability of acquisition of elite sports achievement,' he writes, 'failed to reveal reproducible evidence for any genetic constraints for attaining elite levels by healthy individuals (excluding, of course, the evidence on body size).'

Rather, nonachievers seem to be missing something in their *process* – one or more aspects of style or intensity of practice, or technique, or mindset, or response to failure.

Genes are involved, of course. They're a dynamic part of the process as they become activated. 'When individuals deliberately push themselves beyond the zone of relative comfort and engage in sustained strenuous physical activity,' Ericsson explains, 'they [induce] an abnormal state for cells in some physiological systems . . . These biochemical states will trigger the activation [of] dormant genes within the cells' DNA. The activated genes in turn will stimulate and "turn on" systems designed to cause bodily reorganisation and adaptive change.'

The exact same thing happens with any sustained intellectual or creative activity – chess, for example. As it does for every London cabbie, the brain will physically adapt to any intellectual stretch its owner demands.

All of this nicely reinforces the original double lesson from Ericsson's original 1980 memory experiment: there is no escaping basic human biology – nor is there any need to. Becoming great at something requires the right combination of resources, mentality, strategies, persistence, and time; these are tools theoretically available to any normal functioning human being. This does not mean, of course, that every person has the same resources and opportunity, or that anyone can be great at anything; biological and circumstantial differences and advantages/disadvantages abound. But in revealing talent to be a process, the simple idea of genetic giftedness is forever debunked. It is no longer reasonable to attribute talent or success to a specific gene or any other mysterious gift. The real gift, it turns out, belongs to virtually all of us: it is the plasticity and the extraordinary responsiveness built right into basic human biology. The real gift is the GxE dynamic.

The physiology of this process also requires extraordinary amounts of elapsed time – not just hours and hours of deliberate practice each day, Ericsson found, but also thousands of hours over the course of many years. Interestingly, a number of separate studies have turned up the same common number, concluding that truly outstanding skill in any domain is rarely achieved in less than ten thousand hours of practice over ten years' time (which comes to an average of three hours per day). From sublime pianists to unusually profound physicists, researchers have been very hard-pressed to find any examples of truly extraordinary performers in any field who reached the top of their game before that ten-thousand-hour mark.[3]

In fact, contrary to long-standing myth, Mozart's own career fits beautifully with this new insight. A precocious but by no means adult-level musician as a young boy, Mozart's true greatness as a composer developed slowly and steadily over time. 'People make a great mistake who think that my art has come easily to me,' Mozart himself once wrote to his father, as if to make this precise point. 'Nobody has devoted so much time and thought to composition as I.'

As impressive as it was that little Amadeus attempted to compose at a very early age, his early work was far from extraordinary. In reality, his earliest compositions were mere imitations of other composers. His first seven piano concertos, written from ages eleven to sixteen, 'contain almost nothing original,' reports Temple University's Robert Weisberg, and 'perhaps should not even be labelled as being by Mozart.' He was essentially arranging the works of others for performance on the piano and other instruments.

Over about ten years, Mozart voraciously incorporated different styles and motifs and developed his own voice. Critics consider his Symphony no. 29, written ten years after his first symphony, to be his first work of real stature. His first great piano

[3] This ten-thousand-hour phenomenon has recently attracted significant media attention and has become corrupted and confused. Critics have somehow understood it to be a claim that anyone can achieve anything by putting in ten thousand hours of practice. No serious researcher in expertise studies has ever made any such claim. Ericsson and others have merely observed that approximately ten thousand hours of deliberate practice seems to be one of the necessary components to extraordinary achievement.

concerto is widely considered to be the no. 9, 'Jeunehomme', written at age twenty-one. It was his 271st completed composition. *Idomeneo*, his first operatic masterpiece, written three years later, was his thirteenth opera. The most notable thing about his teenage years is not the quality of his work, but his breathtaking output. Given that, the quality seemed to – in due course – take care of itself. Looking at Mozart's works chronologically, there is a clear trajectory of increasing originality and importance leading up to his final three symphonies, written at age thirty-two, which are generally considered his greatest.

Who else has the potential to scale such heights?

Conventional nature-versus-nurture wisdom says very few people, but the clear and exciting lesson from GxE and from Anders Ericsson's research is this: *no one knows*. We do not – and cannot – know our own limits unless and until we push ourselves to them. Finding one's true natural limit in any field takes many years and many thousands of hours of intense pursuit.

What are your limits?

CHAPTER FOUR

The Similarities and Dissimilarities of Twins

Identical twins often do have striking similarities, but for reasons far beyond their genetic profiles. They can also have surprising (and often overlooked) differences. Twins are fascinating products of the interaction between genes and environment; this has been missed as 'heritability' studies have been wildly misinterpreted. In reality, twin studies do not reveal any percentage of direct genetic influence and tell us absolutely nothing about individual potential.

After nineteen captivating seasons with the Boston Red Sox, Ted Williams retired from baseball on September 28, 1960, at age forty-two. To begin with, it was a blessed anniversary: on the same day in 1941, the Kid had gone six for eight in a doubleheader and earned his legendary 0.406 batting average for the season. Now, two decades later, in the eighth inning of his final game, in his last career at-bat, with a stiff neck and other infirmities, Williams stepped to the plate in Fenway Park, swung steady, and ripped one into deep right centre field for a home run. The Red Sox won the game 5–4.

Would there ever be another hitter like him? When Williams died in 2002, at age eighty-three, his son, John Henry, became convinced that his father's particular genius could be equalled only by a perfect replica: a clone. 'Wouldn't it be interesting if in 50 years, we could bring dad back,' John Henry remarked to his half sister, Bobby-Jo. 'What if we could sell dad's DNA and there could be little Ted Williamses all over the world?' Against Bobby-Jo's wishes, John Henry had Ted's body shipped to a cryonics lab in Scottsdale, Arizona, to be frozen and preserved indefinitely at –321 degrees Fahrenheit. 'There will only be one Ted Williams,' ESPN reported cheekily, ' – for now.'

A perfect copy. Even nonexperts intuitively knew it would never be possible to re-create Ted Williams quirk for quirk and swing for swing. Genes aside, Williams had – like all of us – lived a life, made choices and mistakes, enjoyed friendships and endured hardships, collected memories. A clone would make different mistakes and collect different memories; he'd lead a different life.

He'd also have a very different GxE landscape – with an untold number of different gene-environment interactions than his clonal twin. This is the great unexplained truth about clones: the degree to which the GxE dynamic guarantees enormous differences between originals and their copies. Ever since Dolly the sheep, the world has discussed clones as though they are perfect reproductions of adult beings. GxE guarantees that this isn't so.

Take Rainbow the cat and her clone Cc (short for 'Carbon copy'). In 2001, Rainbow became the first household pet to be successfully cloned. Her clone Cc, created and verified by geneticists at Texas A&M University, shares exactly the same nuclear DNA. But she didn't turn out to be much of a carbon copy. The cats look very different, with different-coloured fur (Rainbow sporting the typical calico colours of brown, tan, white, and gold, while Cc is white and grey) and different physiques (Rainbow is plump, while Cc is slender).

They also have different personalities, according to firsthand observers. Rainbow is quiet and calm, while Cc is curious and playful. Even given their age difference, these genetic clones are clearly far from perfect copies of each other. 'Sure, you can clone your favourite cat,' concludes the Associated Press's Kristen Hays. 'But the copy will not necessarily act or even look like the original.'

That is exactly what thoughtful analysts of human cloning have come to realise as well. 'Identical genes don't produce identical people, as anyone acquainted with identical twins can tell you,' write Wray Herbert, Jeffrey Sheler, and Traci Watson in *US News & World Report*. 'In fact, twins are more alike than clones would be, since they have at least shared the uterine environment, are usually raised in the same family, and so forth . . . All the evidence suggests that the two [clones] would have very different personalities.'

Despite this straightforward understanding, many in the media still produced knee-jerk responses based on the old gene-gift paradigm. In their Ted Williams clone story, ESPN found a biologist, Dr. Lee Silver, who said that a Williams clone would have a leg up on everyone else. 'In theory, you could create someone who would be a step ahead of other people,' Silver said. Even if he didn't take full advantage of his special gene-given talent, Silver explained, '[he] might be just a regular major-league player.'

With such misleading rhetoric still coming from some scientists, how could anyone expect the public to understand genes any better? Practically every word reported in the press supports the notion that genes guarantee every person certain baseline attributes. Ted Williams had superior baseball genes, Isaac Stern had superior music genes, and you – well, you have rather ordinary genes. Accept it.

This impression has been heavily reinforced by the extraordinary coverage of reunited identical twins – beginning in modern times with the amazing Jim twins.

In February 1979, in southwestern Ohio, a thirty-nine-year-old man named Jim Lewis tracked down and introduced himself to his long-lost identical twin brother, Jim Springer. For both men, it was like interacting with a living mirror. Not only did they look the same and talk the same, it turns out their lives uncannily resembled each other's. They had each married and divorced a woman named Linda, and then each married a woman named Betty. They each had an adopted brother named Larry and childhood dog named Toy. They had named their respective firstborn children James Alan Lewis and James Allen Springer. They each drank Miller Lite, chain-smoked Salem cigarettes, enjoyed carpentry and mechanical drawing, chewed their nails, suffered migraines, and had served as part-time sheriffs in their respective towns. They both liked maths and disliked spelling in school. They drove the same model and colour car, lived in the same region of Ohio, and had unknowingly vacationed on the same beach in Florida. They were each six feet tall and weighed about 180 pounds.

Like all identical (or monozygotic) twins, Jim and Jim had been born from twin embryos derived from the same fertilised egg. Their single mother put them both up for adoption after birth, and they were separated into different adoptive families at four weeks of age. Coincidentally, they'd been given the same first name by their adoptive parents. One Jim learned he was an identical twin at age eight. The other didn't know until he met his twin.

A reporter somehow picked up the story and published it in the *Minneapolis Tribune*, which caught the attention of University of Minnesota psychologist Thomas Bouchard. Fascinated, Bouchard invited the Jims to campus for a formal investigation. 'I thought we were going to do a single case study,' Bouchard later recalled. '[But then] we got quite a bit of publicity. *People* magazine ran a story. They were on the Johnny Carson show. They really fascinated everybody. And so I wrote a grant proposal.' The grant money poured in, and more separated twins came to the surface. Within a year, Bouchard and colleagues had studied fifteen other pairs; similar research blossomed around the world.

It was one of Charles Darwin's great puzzlements. 'Nothing seems to me more curious,' he once wrote, 'than the similarity and dissimilarity of twins.' How could some identical twins be so alike and others turn out so different? In separated twins, research psychologists like Bouchard thought they had a unique opportunity to find out, a Darwinist's dream shot at distinguishing nature from nurture. Their method was to compare the ratio of similarities/differences in separated identical twins with the same ratio in separated fraternal twins. Since identical twins were thought to share 100 per cent of their DNA and fraternal twins share, on average, 50 per cent of their genetic material (like any ordinary siblings), comparing these two unusual groups allowed for a very tidy statistical calculation.

The end product was an arcane statistical estimate that researchers unfortunately chose to call 'heritability'.

Heritability isn't at all what it sounds like. The word does not mean anything remotely close to the word 'inherited'. As a result of this dreadfully irresponsible word choice, science journalists and the rest of us have been given a profound misimpression of twin studies and what they've proven. Journalists were understandably blown away when Bouchard and colleagues published data that seemed to demonstrate that genes were responsible for roughly:

60 per cent of intelligence
60 per cent of personality
40 to 66 per cent of motor skills
21 per cent of creativity

Such striking statistics, combined with the compelling stories of the Jims and other sets of twins, had a tidal-wave effect on the media and other scientists. Tragically (predictably), 'heritable' and 'inherited' quickly became interchangeable in the popular lexicon, which led to absurdly reductionist statements like these:

'Since personality is heritable . . .' (*New York Times*)
'In some significant ways . . . criminals are born, not made.' (Associated Press)
'Men's Fidelity Controlled by "Cheating Genetics"' (Drudge Report)

In his 1997 book *Twins*, the award-winning journalist Lawrence Wright extolled what he and others saw as Bouchard's breathtaking scientific achievement. Wright went so far as to declare that Francis Galton and the genetic determinists had been right all along:

'The genetic idea has had a tumultuous passage through the twentieth century,' he wrote, 'but the prevailing view of human nature at the end of the century resembles in many ways the view we had at the beginning . . . Circumstances do not so much dictate the outcome of a person's life as they reflect the inner nature of the person living it. Twins have been used to prove a point, and the point is that we don't become. We are.'

Alas, Wright and other well-meaning journalists relying on Bouchard got it very wrong. Not understanding the true meaning of 'heritability', nor the importance of gene-environment interaction, they radically overstated the direct influence of genes. To be sure, twin studies did – unequivocally – prove that genes are an important and constant influence. Around the world, researchers were able to replicate the basic finding that identical twins had a higher concordance than fraternal twins of intellect, character, and just about everything else. This certainly helped put to rest old arguments from the past that every individual is a blank slate completely shaped by his or her environment.

Blank Slate is dead. Genetic differences do matter.

But the nature of that genetic influence is easily – and perilously – misinterpreted. If we are to take the word 'heritability' at face value, genetic influence is a powerful direct force that leaves individuals rather little wiggle room. Through the lens of this word, twin studies reveal that intelligence is 60 per cent 'heritable', which implies that 60 per cent of each person's intelligence comes preset from genes while the remaining 40 per cent gets shaped by the environment. This appears to prove that our genes control much of our intelligence; there's no escaping it.

In fact, that's not what these studies are saying at all.

Instead, twin studies report, on average, a statistically detectable genetic influence of 60 per cent. Some studies report more, some a lot less. In 2003, examining only poor families, University of Virginia psychologist Eric Turkheimer found that intelligence was not 60 per cent heritable, nor 40 per cent, nor 20 per cent, but *near* 0 per cent – demonstrating once and for all that there is no set portion of genetic influence on intelligence. 'These findings,' wrote Turkheimer, 'suggest that a model of [genes plus environment] is too simple for the dynamic interaction of genes and real-world environments during development.'

How could the number vary so much from group to group? This is how statistics work. Every group is different; every heritability study is a snapshot from a specific time and place, and reflects only the limited data being measured (and how it is measured).

More important, though, is that all of these numbers pertain *only* to groups – not to individuals. Heritability, explains author Matt Ridley, 'is a population average, meaningless for any individual person: you cannot say that Hermia has more heritable intelligence than Helena. When somebody says that heritability of height is 90 per cent, he does not and cannot mean that 90 per cent of my inches come from genes and 10 per cent from my food. He means that variation in a particular sample is attributable to 90 per cent genes and 10 per cent environment. There is no heritability in height for the individual.'

This distinction between group and individual is night and day. No marathon runner would calculate her own race time by averaging the race times of ten thousand other runners; knowing the average lifespan doesn't tell me how long my life will be; no one can know how many kids you will have based on the national average. Averages are averages – they are very useful in some ways and utterly useless in others. It's useful to know that genes matter, but it's just as important to realise that twin studies tell us nothing about you and your individual potential. No group average will ever offer any guidance about individual capability.

In other words, there's nothing wrong with the twin studies themselves. What's wrong is associating them with the word 'heritability', which, as Patrick Bateson says, conveys 'the extraordinary assumption that genetic and environmental influences are independent of one another and do not interact. That assumption is clearly wrong.' In the end, by parroting a strict 'nature vs. nurture' sensibility, heritability estimates are statistical phantoms; they detect something in populations that simply does not exist in actual biology. It's as if someone tried to determine what percentage of the brilliance of *King Lear* comes from adjectives. Just because there are fancy methods available for inferring distinct numbers doesn't mean that those numbers have the meaning that some would wish for.

So what about Darwin's essential question: How could some identical twins be so alike and others turn out so different? Leaving behind the misconceptions of heritability, developmental biologists and psychologists suggest the following real-world considerations for why twins turn out the way they do:

1. **Early shared GxE.** Identical twins share a wide collection of similarities not just because they share the same genes, but because they share the same genes *and* early environments – hence, the same gene-environment interactions throughout gestation.

2. **Shared cultural circumstances.** In identical-twin comparisons, shared biology always grabs all the attention. Inevitably overlooked is the vast number of shared cultural traits: same age, same sex, same ethnicity, and, in most cases, a raft of other shared (or very similar) social, economic, and cultural experiences. 'All of these factors work towards increasing the resemblance of reared-apart twins,' explains psychologist Jay Joseph.

Just how powerful are these shared cultural influences? To test the influence of just a few of them, psychologist W. J. Wyatt assembled fifty college students completely unrelated and unknown to one another and then placed them in random pairings purely on the basis of age and sex. Among the twenty-five pairs, one pair showed a remarkable collection of similarities: both Baptists, both pursuing nursing careers, both passionate about volleyball and tennis, both favoured English and maths, both detested shorthand, and both preferred to vacation at historical places. The point from this very limited study is not to draw definitive conclusions about specific environmental influences, but to draw attention to the power of unseen similar circumstances.

3. **Hidden dissimilarities.** Statisticians call it 'the multiple-end-point problem': the seductive trap of selectively picking data

that fit a certain thesis, while conveniently discarding the rest. For every tiny similarity between the Jim twins, there were thousands of tiny (but unmentioned) dissimilarities. 'There are endless possibilities for doing bad statistical inferences,' says Stanford statistician Persi Diaconis. 'You get to pick which features you want to resonate to. When you look at your mom, you might say, "I'm exactly the opposite." Someone else might say, "Hmm."'

New York Times science writer Natalie Angier adds, 'What the public doesn't hear of are the many discrepancies between the twins. I know of two cases in which television producers tried to do documentaries about identical twins reared apart but then found the twins so distinctive in personal style – one talky and outgoing, the other shy and insecure – that the shows collapsed of their own unpersuasiveness.'

4. **Coordination and exaggeration.** All twins feel a close bond with each other, and while child twins growing up together might often cling to their differences, reunited adult twins understandably revel in their similarities. Researchers try to guard against any purposeful or unwitting coordination, but in her 1981 book *Identical Twins Reared Apart*, Susan Farber reviewed 121 cases of twins described by researchers as 'separated at birth' or 'reared apart'. Only three of those pairs had actually been separated shortly after birth *and* studied at their first reunion. At the University of Minnesota, the average age of separated twins studied turned out to be forty, while their average years spent apart was thirty – leaving an average of ten years of contact prior to research interviews.

Considering all this, was it really so shocking that Jim Lewis and Jim Springer, two thirty-nine-year-old men who shared a womb for nine months and a month more in the same hospital room, *and* were raised in working-class towns seventy miles apart (by parents with tastes similar enough to name their kids Jim and Larry), would end up preferring the same beer, same cigarettes, same car, same hobbies, and have some of the same habits? (Lest anyone think they were living perfectly parallel lives, it's also worth noting a few of their differences: One of the Jims was married a third time. They wore their hair very differently. One was much more verbally articulate than the other . . .)

By the same token, should it really surprise anyone to see the picture of identical twins below? Otto (left) and Ewald (right), age twenty-three, had trained intensively for different athletic advantages – Otto as a long-distance runner, and Ewald for strength events.

Gene-gift proponents would have us believe in genetic fate and strict limits. The real lesson of genetics, though – and of identical twins – is precisely the opposite. None of us is stuck in some sort of destined body or life. We inherit – and we also become.

CHAPTER FIVE

Prodigies and Late Bloomers

Child prodigies and superlative adult achievers are often not the same people. Understanding what makes remarkable abilities appear at different phases of a person's life provides an important insight into what talent really is.

In Michael Jordan's prime, he could leap so far to the hoop, and remain airborne for so long, it looked as though he could actually defy gravity. They called it 'hang time' – that spectacular second or two during which Jordan seemed to suspend himself in midair and fly forward, rolling his tongue, pumping his wing-legs, and finally stuffing the ball. Then he'd gently descend back to earth. It was far from the only move in his arsenal; for several years Jordan could move, shoot, pass, defend, and dunk so much better than any other player that he took on a superhuman aura. Near the end of his career, when Jordan confided to Chicago Bulls coach Phil Jackson his intention to retire, the coach responded with an unusual appeal. 'Michael,' he said, 'pure genius is something very, very rare and if you are blessed enough to possess it, you want to think a long time before you walk away from using it.'

But where did that indisputable 'pure genius' come from? Interestingly, it had been nowhere in sight during Jordan's childhood. Michael was not the best athlete in the Jordan family as a youth (his older brother Larry was); not the most industrious (of five siblings, he was by far the laziest); and not very mechanically inclined (a prized family skill). 'If Michael Jordan was some kind of genius, there had been few signs of it when he was young,' writes David Halberstam in his biography *Playing for Keeps*. In his sophomore year of high school (age 15–16), after attending summer basketball camp with his friend Roy Smith, Jordan didn't even make the varsity basketball squad. Smith did.

The virtuoso cellist Yo-Yo Ma, on the other hand, showed his stuff from very early on, dazzling his piano teacher at age three, playing difficult work by Bach on the cello at age five, and performing for Leonard Bernstein and President John F. Kennedy at age seven. On hearing the young Yo-Yo play for the first time, legendary cellist Pablo Casals called him simply 'Wonder Boy'.

What makes abilities come into view at such different times in a person's life? In the popular imagination, a person either has talent or does not; if so, it flows through him or her like an invisible river of energy, constant and timeless. The reality, though, is that even superachievers develop very different abilities at different ages – so much so, in fact, that researchers have discovered that child prodigies and adult superachievers are very often not the same people. For every wonder child like Yo-Yo Ma who also thrives in adulthood, there is a long list of child prodigies who never become remarkable adult achievers. At the same time, an equally long list of profound adult achievers manage to attain greatness without first showing any profound abilities as children – a list that includes Copernicus, Rembrandt, Bach, Newton, Kant, da Vinci, and Einstein.

Only one paradigm – talent as process – can make sense of all these great achievements from such radically different life stages. All individuals have distinct biologies, but no one has a predestined biological fate. Every individual is built with the capacity, as Patrick Bateson says, 'to develop in a number of distinctly different ways.' To discover your own potential, add water, love, perseverance, and lots and lots of time.

Unfortunately, some talent researchers still insist on

categorising causes as either nature or nurture, describing them as additive (G+E) rather than interactive (GxE), and presenting core abilities as innate and immutable – when contemporary science so plainly points to a more interactive dynamic.

It is a difficult legacy to shed, given what seems like such clear evidence for innate talent right in front of our eyes. Without question, child prodigies exist and always have. The eighteenth-century English jurist Jeremy Bentham began studying Latin at age three and entered Oxford University at age twelve. The mathematician John von Neumann could divide eight-digit numbers in his head by age six. Hungary's Judit Polgár became a chess grandmaster at age fifteen. Seattle's Adora Svitak began writing stories at age five and published her first book at age seven. Over many centuries, we have reliable records of young children demonstrating extraordinary abilities in maths, music, language, spatial intelligence, and the visual arts.

Where do these extraordinary abilities come from? Because they appear so early (parents often say 'out of nowhere') and are often so bewitching, the most common instinct from parents and researchers alike is to answer the big mystery with a simple idea: such talent is an inborn gift. In the 1990s, Anders Ericsson and others challenged that long-held view by bringing the talent-formation process partially into the light, documenting a new 'science of high ability'. In the face of Ericsson's paradigm-challenging data, though, other researchers pushed back. Boston College's Ellen Winner responded in 2000 that 'Ericsson's research demonstrates the importance of hard work but does not rule out the role of innate ability . . . [We] conclude that intensive training is necessary for the acquisition of expertise, but *not* that it is sufficient.' An exceptional 'inborn giftedness' must also be present, she argued.

'Necessary but not sufficient' became a common reaction to Ericsson as many professionals clung to the unsustainable notion of innate gifts. This critique overlooked the possibility of an entirely new model that recognised training and biology as one interconnected, dynamic force.

Driving Winner's argument were two core beliefs:

1. Some extraordinary abilities appear earlier than they could possibly be developed.
2. There is evidence for what she called 'atypical brain organisation' in gifted children, occurring 'as a result of genetics, the in-utero environment, or after-birth trauma.'

Her first point has, historically, been the most popular driver of the giftedness paradigm: since one cannot see talent being developed, it must simply exist. But is this thinking still justified, given what we've learned? As noted in earlier chapters, studies have now shown conclusively that mind-set, nutrition, parenting, peers, media culture, time, focus, and motivation all profoundly affect the development of abilities. All of these factors are in play from the first day of a child's life (or earlier). We need look no further than Hart and Risley's spoken-word study to understand how early life experience dramatically affects the trajectory of a very young child. We also know for sure that early musical exposure can work the same way. The same experience has been documented with chess players. Like any taxi driver's brain, a young child's brain adapts to demands. The process is very slow and impossible to see from the outside, but it still happens. Imperceptibly, like water evaporating into a rain cloud, tiny events pave the way for development in one direction or another.

As to Winner's second point, it is indisputably true that some people with extraordinary abilities have distinct physiological differences in their brains. For example, Winner points out that mathematically and musically 'gifted' individuals tend to use both lobes of the brain for tasks usually dominated by the left hemisphere in individuals with normal abilities, and that artists, inventors, and musicians tend to have a higher proportion of language disorders. But does it follow that these differences are innate? Winner's list of three possible causes – genetics, in utero environment, and after-birth trauma – actually all turn out to be dynamic actors in every person's development. Consider that 'genetics' actually means 'genetic expression' and that the uterine environment and after-birth events are both highly developmental, and the notion of 'innate' quickly dissolves away. Further, there's no logical reason why her list should be limited to three possible causes. If she's allowing for after-birth trauma, why not also allow for other infant and toddler life experiences?

The very rare phenomenon of spectacular savants like Kim Peek (the 'real Rain Man') points even more clearly to developmental dynamics rather than hardwired abilities. Peek, who died in late 2009 at the age of 58, was severely cognitively disabled, could not button his own shirt, and tested very poorly on a standard IQ test – but had memorised many thousands of books word for word. He was one of an estimated one hundred contemporary prodigious savants with both severe impairments and extraordinary abilities. The group also includes Daniel Tammet, who lives with autism but who can recite pi to 22,514 digits and was able to add Icelandic to his other nine languages in just nine days; Leslie Lemke, who couldn't stand until twelve years of age or walk until fifteen – but one night at age sixteen began playing every note of Tchaikovsky's Piano Concerto no. 1 after hearing it just once on television; and Alonzo Clemons, who, ever since a childhood head injury, has been unable to feed himself or tie his own shoes but can sculpt an animal in exquisite detail after seeing a fleeting image of it.

University of Wisconsin psychiatrist Darold Treffert, perhaps the world's leading expert on what he calls 'savant syndrome', points out that these are actually extreme examples of a more universal phenomenon. He estimates that approximately one in ten persons with autism has some savant skills. The syndrome, he explains, occurs when the brain's left hemisphere is severely damaged, inviting the right hemisphere (which is responsible for things like music and art) to compensate heavily for the loss.

It is critical to note that the damage does not create the ability; rather, it creates the opportunity for the ability to develop. This, says Treffert, 'promote[s] the idea of the brain's plasticity, and the brain's ability to recruit other areas to be put to use.'

In fact, it has prompted Treffert to wonder aloud, 'Might there be a little Rain Man in each of us?'

> In the case of the prodigious savant, it appears to me, there is a marvellous coalescence of idiosyncratic brain circuitry [combined with] obsessive traits of concentration & repetition and tremendous encouragement & reinforcement from family, caretakers and teachers. Does some of that same possibility, a little Rain Man as it were, perhaps reside within each of us? I think that it does.

Other savant researchers heartily agree. In 2003, the University of Sydney's Allan W. Snyder and colleagues used magnetic pulses to temporarily impair the left frontotemporal lobe in healthy persons, resulting in some temporary savant-like tendencies – drawing animals with more detail, for example, and proofreading with more accuracy. Shutting off portions of their brain did not suddenly transform them into amazing artists or brilliant thinkers; rather, it altered their way of thinking and observing, shifting attention away from meaning and understanding and toward detail. Such an effect, Snyder and colleagues noted, can be achieved through other means. 'Apart from brain impairment and magnetic stimulation,' they wrote, 'savant-like skills might also be made accessible by altered states of perception or by EEG-assisted feedback. [Oliver] Sacks provides support for the former view. He produced camera-like precise drawings only when under the influence of amphetamines. Early (savant-like) cave art has been attributed to mescaline induced perceptual states.'

Even very ordinary brains are capable of extraordinary things when provoked.

. . .

Perhaps the most interesting longitudinal study on giftedness comes from IQ inventor and staunch innate-intelligence advocate Lewis Terman (last mentioned in chapter 2). In the early 1920s, Terman began a massive, decades-long study of child achievers, which he pointedly called 'Genetic Studies of Genius'. It was his contention that the most successful children were endowed with elite genes propelling them to lifelong success. To prove this, he began tracking nearly fifteen hundred California schoolkids identified as 'exceptionally superior'. Alas, as Terman's exceptional kids matured, they seemed less and less exceptional. They did grow up to be healthier and more successful than the average American, but very few ultimately emerged as geniuses or superachievers. None went on to earn the Nobel Prize – as two children *rejected* from Terman's original group did. None became world-class musicians – as two other Terman rejects, Isaac Stern and Yehudi Menuhin, did. All in all, Terman's epic studies in genius turned out to be studies in disappointment.

The frustration was especially keen when it came to the very top of Terman's group – the 5 per cent who had scored 180 or better on the IQ test. 'One is left with the feeling that the above-180 IQ subjects were not as remarkable as might have been expected,' concluded Tufts's David Henry Feldman in a 1984 retrospective of the long study. 'There is the disappointing sense that they might have done more with their lives.'

A few years later, Feldman concluded his own separate study of six child prodigies in music, art, chess, and maths. None of his subjects grew up to become extraordinary adult achievers. In her research, Ellen Winner had found exactly the same thing. 'Most gifted children, even most child prodigies, do not go on to become adult creators,' she reported.

Why?

First off, it turns out that the skill sets are very different. The attributes necessary for high child achievement are simply not the same as those that drive adult achievement, so one would not automatically flow from the other. 'A high IQ six-year-old who can multiply three-digit numbers in her head, or solve algebraic equations, wins acclaim,' explains Winner. 'But as a young adult, she must come up with some new way to solve some unsolved mathematical problem, or must discover some new problems or areas to investigate. Otherwise, she will not make her mark in the domain of mathematics . . . The situation is the same in art or music. Technical perfection wins the prodigy adoration, but if the prodigy does not eventually go beyond this, he or she sinks into oblivion.'

The second reason is even more interesting: child achievers are frequently hobbled by the psychology of their own success. Children who grow up surrounded by praise for being technically proficient at a specific task often develop a natural aversion to stepping outside their comfort zone. Instead of falling into a pattern of taking risks and regularly pushing themselves just beyond their limit, they develop a terrible fear of new challenges and of any sort of flaw or failure. Ironically, this leads them away from the very building blocks of adult success. 'Prodigies [can] become frozen into expertise,' says Ellen Winner. 'This is particularly a problem for those whose work has become public and has won them acclaim, such as musical performers, painters, or children who have been publicised as "whiz kids" . . . It is difficult to break away from [technical] expertise and take the kinds of risks required to be creative.'

Underneath all this is the core reality that talented children and their parents frequently do not notice the development of their own skills during infancy and toddlerhood. This is perfectly understandable – obviously, tiny children themselves can't notice such things, and for parents to take note of such a nuanced process in fine detail could be construed as odd and obsessive – but it can also lead to a grave logical error: failure to see it as a process may inspire the conclusion that a collection of skills is really an innate gift. 'Mommy, I don't know,' the toddler Yo-Yo Ma replied to his mother, Marina, when she asked how he could sense an out-of-tune note. 'I just *know*.' What was the true source of Yo-Yo's uncanny ability? In her memoir, his mother chalks it up to genetics – but then she details how, from the very moment of his birth, Yo-Yo was exposed to music in the most profound and exquisite way. Both Marina, a trained opera singer, and her husband, Hiao-Tsiun, a teacher/composer/conductor, had immigrated to Paris as young adults to study, play, compose, and teach music. Having traded comfort and status in China for immigrant poverty in France, the Ma family breathed, ate, and slept music. Their tiny two-room Paris apartment was arranged thusly:

> Mother and children slept in one room; the other, a smaller bedroom-studio, was used by Hiao-Tsiun. Amazingly he had squeezed into that room his piano, a collection of children's string instruments, and his cot. His precious manuscripts and music scores, meticulously arranged by him for children, were jammed into an old armoire and piled up on the piano top. Every corner was bulging with his papers.

Hiao-Tsiun studied at the conservatory by day and gave lessons in the evening, all the while clinging to his deeply personal dream of creating a children's orchestra. Like Leopold Mozart, he designed elaborate pedagogical techniques specifically for children and was eager to put them to use. Yo-Yo's older sister, Yeou-Cheng, was (like Nannerl Mozart) started on piano and violin at a very early age – around the time Yo-Yo was born. By the time Yo-Yo was ready to start piano at age three, his sister was already a budding prodigy. 'From the cradle, Yo-Yo was surrounded by a world of music,' his mother recalls. 'He heard hundreds of classical selections on records, or played by his father or his sister. Bach and Mozart were engraved on his mind.'

Engraved on his mind: according to neuroscientists and music psychologists, this is quite literally true. We know now that music activates neurons in many regions of the brain simultaneously and that every meaningful listening experience inspires the formation of multiple-trace memories, which, in turn, inform the encoding of all future musical memories. 'Melodic "calculation centres" in the dorsal temporal lobes appear to be paying attention to interval size and distances between pitches as we listen to music, creating a pitch-free template of the very melodic values we will need in order to recognise songs in transposition,' explains McGill University's Daniel Levitin. Levitin also concurs with University of California, San Diego's Diana Deutsch and others in deducing that every human being is likely born with the capacity for absolute pitch, but that it gets activated only in those who are exposed to enough tonal imprinting at a very early age.

In addition to the neural mechanics, there were also powerful psychological forces in Yo-Yo Ma's life that helped shape him into an obsessively determined musician at the youngest possible age. Yo-Yo worshipped his sister and father and desperately wanted to impress both. From very early on, he responded to his stern father – who had vowed to 'make a musician of him' at age two – with a blend of admiration, duty, and extreme stubbornness. Yo-Yo would hover in the doorway while his sister practised and, when asked, critique her performances note by note. In his own performances Yo-Yo was determined to have it his way. Sometimes, he refused to perform for his parents as instructed; other times, he would play more than he was supposed to.

He also needed to chart his own course instrumentally. 'I don't like the sound violins make,' Yo-Yo informed his father at age four. 'I want a big instrument.'

'Once you start playing with a big instrument, you cannot switch back to the violin,' Hiao-Tsiun responded firmly to his four-year-old son. 'Don't tell me a month from now that you have changed your mind.'

'I will play it,' Yo-Yo insisted. 'I won't change my mind.'

And he did not. In retrospect, his early life contained all the known ingredients for the brewing of extraordinary achievement: an early and intensively conditioned musical brain, world-class teaching resources, and a desperate personal desire that researchers universally agree is *the* key to precocious success. Ellen Winner calls it 'the rage to master', a fervent, never-let-go willfulness and focus that drives a child into an early version of Ericsson's deliberate practice.

As a general rule, high achievers have exceptional drive. From Olympic athletes to Nobel physicists, from long-winded U.S. senators to the shyest poet laureate, you simply don't see remarkable achievement without it. The question is, why does this obsessive need appear at different ages in different people, and why does it not appear in some people at all? If it was simply a matter of genetics, as Lewis Terman proposed, we would indeed see the pattern of lives he imagined with his Genetic Studies of Genius project. Instead, intense ambition evolves out of complex, real-world dynamics, settling into people's psyches at different ages and circumstances – sometimes from extreme adversity, sometimes as a proxy for revenge, sometimes as a way of proving oneself to a beloved/feared parent or sibling, and so on. The collection of potential catalysts for intense ambition may never be entirely understood and will surely never be easily reproducible. But that doesn't mean we shouldn't try to understand the mechanism better, or apply its lessons.

Michael Jordan always seemed to hate losing (an everyday experience while growing up with his brother Larry), but his willingness to do absolutely anything to improve his skills didn't appear until after his rejection from the varsity squad in tenth grade (age 15–16). At that point, according to his friend Roy Smith, his competitiveness went into overdrive. Laney High assistant coach Ron Coley remembers his very first sighting of Jordan, near the end of a junior varsity basketball game that year. 'There were nine players on the court just coasting,' Coley recalls, 'but there was one kid playing his heart out. The way he was playing, I thought his team was down one point with two minutes to play. So I looked up at the clock and his team was down twenty points with one minute to play. It was Michael.'

For the remainder of his basketball career, no one within Jordan's orbit ever practised or played as hard. 'All top athletes are driven,' writes David Halberstam, 'and no one made the [University of North] Carolina roster unless he was by far the hardest-working kid in his neighbourhood, his high school and finally his high school conference, but Jordan was self-evidently the most driven of all.' In a college programme famous for its loyalty and dedication, Jordan impressed Carolina coach Dean Smith with his extra level of ferocity. In fact, he seemed to get more intense with each passing year. As he returned for his sophomore (second) year, fellow players noticed yet another bump in both confidence and zeal. 'Even in pickup games,' writes Halberstam, 'he had become unusually purposeful. There was a tendency in games like this, when there were no coaches around, for players to resort to what they did best, to reinforce their strengths and avoid going to any part of their game that was essentially weak. But Jordan [was] constantly working on the weaker part of his game trying to bring it up. It [was] one more sign of his desire to be the best.' Coach Smith found that, in practice, Jordan was now winning all the one-on-one games and all the five-on-five games. So he started stacking the deck – giving Jordan weaker and weaker teammates to make him work even harder to win. That seemed to spur him on to even further greatness. After his junior year, Smith realised there was nothing else he could do for him, and he pushed Jordan to leave college ball for the NBA.

One common characteristic in all successful adults is that, at some point in their lives, they come to realise how much the process of improvement is within their own control. That's also what Stanford psychologist Carol Dweck observed in a series of grade-school studies in the 1990s. In her central experiment, Dweck (who was then at Columbia) asked four hundred seventh graders (age 12–13) to complete a relatively easy set of puzzles and then randomly separated them into two groups. Individually, each student in the first group was complimented for his or her innate intelligence with the line, 'You must be smart at this!'

Each student in the second group was praised for his or her effort: 'You must have worked really hard!'

Then each child was offered a chance to take one of two follow-up tests: either another easy set of puzzles or a much harder set of puzzles that teachers promised would be a great learning experience.

The results:

- More than half of the kids praised for their inborn intelligence chose the easy follow-up puzzle.
- A staggering 90 per cent of the kids praised for their hard work chose the more difficult puzzles.

Other Dweck experiments pointed in the same direction, demonstrating irrefutably that people who believe in inborn intelligence and talents are less intellectually adventurous and less successful in school. By contrast, people with an 'incremental' theory of intelligence – believing that intelligence is malleable and can be increased through effort – are much more intellectually ambitious and successful.

The lesson is that parents, teachers, and students must take the long and incremental view. Regardless of whether a child seems to be exceptional, mediocre, or even awful at any particular skill at a particular point in time, the potential exists for that person to develop into a high-achieving adult. Because talent is a function of acquired skills rather than innate ability, adult achievement depends completely on long-term attitude and resources and process rather than any particular age-based talent quotient. While childhood achievement is, of course, not irrelevant (it's often a sign of early interest and determination), it doesn't rule any particular future success in or out.

Childhood abilities – or lack thereof – are not a crystal ball of future success. No age-related level of achievement is either a golden ticket or a locked gate.

CHAPTER SIX

Can White Men Jump?

Ethnicity, Genes, Culture, and Success

Clusters of ethnic and geographical athletic success prompt suspicions of hidden genetic advantages. The real advantages are far more nuanced – and less hidden.

At the 2008 Summer Olympics in Beijing, the world watched in astonishment as the tiny island of Jamaica captured six gold medals in track and field and eleven overall. Usain Bolt won (and set world records in) both the men's 100-metre and the men's 200-metre races. Jamaican women took the top three spots in the 100-metre and won the 200-metre as well. 'They brought their A game. I don't know where we left ours,' lamented American relay runner Lauryn Williams.

A poor, underdeveloped nation of 2.8 million people – one-hundredth the size of the United States – had somehow managed to produce the fastest humans alive.

How?

Within hours, geneticists and science journalists rushed in with reports of a 'secret weapon': biologically, it turned out that almost all Jamaicans are flush with alpha-actinin-3, a protein that drives forceful, speedy muscle contractions. The powerful protein is produced by a special gene variant called *ACTN3*, at least one copy of which can be found in 98 per cent of Jamaicans – far higher than in many other ethnic populations.

An impressive fact, but no one stopped to do the maths. Eighty per cent of Americans also have at least one copy of *ACTN3* – that amounts to 240 million people. Eighty-two per cent of Europeans have it as well – that tacks on another 597 million potential sprinters. 'There's simply no clear relationship between the frequency of this variant in a population and its capacity to produce sprinting superstars,' concluded geneticist Daniel MacArthur.

What, then, is the Jamaicans' secret sauce?

This is the same question people asked about champion long-distance runners from Finland in the 1920s and about great Jewish basketball players from the ghettos of Philadelphia and New York in the 1930s. Today, we wonder how tiny South Korea turns out as many great female golfers as the United States – and how the Dominican Republic has become a factory for male baseball players.

The list goes on and on. It turns out that sports excellence commonly emerges in geographic clusters – so commonly, in fact, that a small academic discipline called 'sports geography' has developed over the years to help understand it. What they've discovered is that there's never a single cause for a sports cluster. Rather, the success comes from many contributions of climate, media, demographics, nutrition, politics, training, spirituality, education, economics, and folklore. In short, athletic clusters are not genetic, but systemic.

Unsatisfied with this multifaceted explanation, some sports geographers have also transformed themselves into sports geneticists. In his book *Taboo: Why Black Athletes Dominate Sports and Why We're Afraid to Talk About It*, journalist Jon Entine insists that today's phenomenal black athletes – Jamaican sprinters, Kenyan marathoners, African American basketball players, etc. – are propelled by 'high performance genes' inherited from their West and East African ancestors. Caucasians and Asians don't do as well, he says, because they don't share these advantages. 'White athletes appear to have a physique between central West Africans and East

Africans,' Entine writes. 'They have more endurance but less explosive running and jumping ability than West Africans; they tend to be quicker than East Africans but have less endurance.'

In the finer print, Entine acknowledges that these are all grosser-than-gross generalisations. He understands that there are extraordinary Asian and Caucasian athletes in basketball, running, swimming, jumping, and cycling. (In fact, blacks do not even dominate the latter three of these sports as of 2008.) In his own book, Entine quotes geneticist Claude Bouchard: 'The key point is that these biological characteristics *are not unique* to either West or East African blacks. These characteristics are seen in all populations, including whites.' (Italics mine.) (Entine also acknowledges that we haven't in fact found the actual genes he's alluding to. 'These genes will likely be identified early in the [twenty-first century],' he predicts.)

Actual proof for his argument is startlingly thin. But Entine's message of superior genes seems irresistible to a world steeped in gene-giftedness – and where other influences and dynamics are nearly invisible.

Take the running Kenyans. Relatively new to international competition, Kenyans have in recent years become overwhelmingly dominant in middle- and long-distance races. 'It's pointless for me to run on the pro circuit,' complained American 10,000-metre champion Mike Mykytok to the *New York Times* in 1998. 'With all the Kenyans, I could set a personal best time, still only place 12th and win $200.'

Ninety per cent of the top-performing Kenyans come from the Kalenjin tribe in the Great Rift Valley region of western Kenya, where they have a centuries-old tradition of long-distance running. Where did this tradition come from? Kenyan-born journalist John Manners suggests it came from cattle raiding. Further, he proposes how a few basic economic incentives became a powerful evolutionary force. 'The better a young man was at raiding [cattle] – in large part a function of his speed and endurance – the more cattle he accumulated,' Manners says. 'And since cattle were what a prospective husband needed to pay for a bride, the more a young man had, the more wives he could buy, and the more children he was likely to father. It is not hard to imagine that such a reproductive advantage might cause a significant shift in a group's genetic makeup over the course of a few centuries.'

Whatever the precise origin, it is true that the Kalenjin have long had a fierce dedication to running. But it wasn't until the 1968 Olympics that they became internationally renowned for their prowess, thanks to the extraordinary runner Kipchoge Keino.

The son of a farmer and ambitious long-distance runner, Keino caught the running bug early in life. He wasn't the most precocious or 'natural' athlete among his peers, but running was simply woven into the fabric of his life: along with his schoolmates, Keino ran many miles per day as a part of his routine. 'I used to run from the farm to school and back,' he recalled. 'We didn't have a water tap in the house, so you run to the river, take your shower, run home, change, [run] to school . . . Everything is running.' Slowly, Keino emerged as a serious competitor. He built himself a running track on the farm where his family worked and by his late teens was showing signs of international-level performance. After some success in the early 1960s, he competed admirably in the 1964 Olympics and became the leader of the Kenyan running team for the 1968 games in Mexico City. It was Kenya's fourth Olympics.

In Mexico City, things did not begin well for Keino. After nearly collapsing in pain during his first race, the 10,000 metres, he was diagnosed with gallstones and ordered by doctors not to continue. At the last minute, though, he stubbornly decided to race the 1,500 metres and hopped in a cab to Mexico City's Aztec Stadium. Caught in terrible traffic, Keino did the only thing he could do, the thing he'd been training his whole life for: he jumped out of the cab and ran the last mile to the event, arriving on the track only moments before the start of the race, winded and very sick. Still, when the gun sounded, Keino was off, and his performance that day shattered the world record and left his rival, American Jim Ryun, in the dust.

The dramatic victory made Keino one of the most celebrated men in all Africa and helped catalyse a new interest in world-class competition. Athletic halls and other venues all over Kenya were named after him. World-class coaches like Fred Hardy and Colm O'Connell were recruited to nurture other Kenyan aspirants. In the decades that followed, the long-standing but profitless Kalenjin running tradition became a well-oiled economic-athletic engine. Sports geographers point to many crucial ingredients in Kenya's competitive surge but no single overriding factor. High-altitude training and mild year-round climate are critical, but equally important is a deeply ingrained culture of asceticism – the postponement of gratification – and an overriding preference for individual over team sports. (Soccer, the overwhelming Kenyan favourite, is all but ignored among the Kalenjin; running is all.) In testing, psychologists discovered a particularly strong cultural 'achievement orientation', defined as the inclination to seek new challenges, attain competence, and strive to outdo others. And then there was the built-in necessity as virtue: as Keino mentioned, Kalenjin kids tend to run long distances as a practical matter, an average of eight to twelve kilometres per day from age seven.

Joke among elite athletes: How can the rest of the world defuse Kenyan running superiority? Answer: Buy them school buses.

With the prospect of international prize money, running in Kenya has also become a rare economic opportunity to catapult oneself into Western-level education and wealth. Five thousand dollars in prize money is a very nice perk for an American; for a Kenyan, it is instant life-changing wealth. Over time, a strong culture of success has also bred even more success. The high-performance benchmark has stoked higher and higher levels of achievement – a positive feedback loop analogous to technological innovation in Silicon Valley, combat skills among Navy SEALs, and talents in other highly successful microcultures. In any competitive arena, the single best way to inspire better performance is to be surrounded by the fiercest possible competitors and a culture of extreme excellence. Success begets success.

There is also an apparent sacrificial quality particular to Kenyan training, wherein coaches can afford to push their athletes to extreme limits in a way that coaches in other parts of the world cannot. *Sports Illustrated*'s Alexander Wolff writes that with a million Kenyan schoolboys running so enthusiastically, 'coaches in Kenya can train their athletes to the outer limits of endurance – up to 150 miles a week – without worrying that their pool of talent will be meaningfully depleted. Even if four out of every five runners break down, the fifth will convert that training into performance.'

And what of genetics? Are Kenyans the possessors of rare endurance genes, as some insist? No one can yet know for sure, but the new understanding of GxE and some emergent truths in genetic testing strongly suggest otherwise, in two important ways:

1. DESPITE APPEARANCES TO THE CONTRARY, RACIAL AND ETHNIC GROUPS ARE *NOT* GENETICALLY DISCRETE.

Skin colour is a great deceiver; actual genetic differences between ethnic and geographic groups are very, very limited. All human beings are descended from the same African ancestors, and it is well established among geneticists that there is roughly ten times more genetic variation within large populations than there is between populations. 'While ancestry is a useful way to classify species (because species are isolated gene pools, most of the time),' explains University of Queensland philosopher of biology John Wilkins, 'it is rarely a good way to classify populations within species . . . [and definitely not] in humans. We move about too much.'

By no stretch of the imagination, then, does any ethnicity or region have an exclusive lock on a particular body type or secret high-performance gene. Body shapes, muscle fibre types, etc., are actually quite varied and scattered, and true athletic potential is widespread and plentiful.

2. GENES DON'T DIRECTLY CAUSE TRAITS; THEY ONLY INFLUENCE THE SYSTEM.

Consistent with other lessons of GxE, the surprising finding of the

$3 billion Human Genome Project is that only in rare instances do specific gene variants directly cause specific traits or diseases. Far more commonly, they merely increase or decrease the likelihood of those traits/diseases. In the words of King's College developmental psychopathologist Michael Rutter, genes are 'probabilistic rather than deterministic.'

As the search for athletic genes continues, therefore, the overwhelming evidence suggests that researchers will instead locate genes prone to certain types of interactions: gene variant A in combination with gene variant B, provoked into expression by X amount of training + Y altitude + Z will to win + a hundred other life variables (coaching, injury rate, etc.), will produce some specific result R. What this means, of course, is that we need to dispense rhetorically with the thick firewall between biology (nature) and training (nurture). The reality of GxE assures that each person's genes interact with his climate, altitude, culture, meals, language, customs, and spirituality – everything – to produce unique life trajectories. Genes play a critical role, but as dynamic instruments, not a fixed blueprint. A seven- or fourteen- or twenty-eight-year-old outfitted with a certain height, shape, muscle-fibre proportion, and so on is not that way merely because of genetic instruction.

. . .

As for John Manners's depiction of cattle-raiding Kenyans becoming genetically selected to be better and better runners over the generations, it's an entertaining theory that fits well with the popular gene-centric view of natural selection. But developmental biologists would point out that you could take exactly the same story line and flip the conclusion on its head: the fastest man earns the most wives and has the most kids – but rather than passing on quickness genes, he passes on crucial external ingredients, such as the knowledge and means to attain maximal nutrition, inspiring stories, the most propitious attitude and habits, access to the best trainers, the most leisure time to pursue training, and so on. This nongenetic aspect of inheritance is often overlooked by genetic determinists: culture, knowledge, attitudes, and environments are also passed on in many different ways.

The case for the hidden performance gene is even further diminished in the matter of Jamaican sprinters, who turn out to be a quite heterogeneous genetic group – nothing like the genetic 'island' that some might imagine. On average, Jamaican genetic heritage is about the same as African American heritage, with roughly the same mix of West African, European, and native American ancestry. That's on average; individually, the percentage of West African origin varies widely, from 46.8 to 97.0 per cent. Jamaicans are therefore *less* genetically African and *more* European and native American than their neighbouring Barbadians and Virgin Islanders. 'Jamaica . . . may represent a "crossroads" within the Caribbean,' conclude the authors of one DNA study. Jamaica was used as a 'transit point by colonists between Central and South America and Europe [which] may have served to make Jamaica more cosmopolitan and thus provided more opportunities for [genetic] admixture to occur. *The large variance in both the global and individual admixture estimates in Jamaica attests to the cosmopolitan nature of the island.*'

In other words, Jamaica would be one of the very last places in the region expected to excel, according to a gene-gift paradigm.

Meanwhile, specific cultural explanations abound for the island's sprinting success – and for its recent competitive surge. In Jamaica, track events are beloved. The annual high school Boys' and Girls' Athletic Championships is as important to Jamaicans as the Super Bowl is to Americans. 'Think Notre Dame football,' write *Sports Illustrated*'s Tim Layden and David Epstein. 'Names like Donald Quarrie and Merlene Ottey are holy on the island. In the United States, track and field is a marginal, niche sport that pops its head out of the sand every four years and occasionally produces a superstar. In Jamaica . . . it's a major sport. When *Sports Illustrated* [recently] visited the island . . . dozens of small children showed up for a Saturday morning youth track practice. That was impressive. That they were all wearing spikes was stunning.'

With that level of intensity baked right into the culture, it's no surprise that Jamaicans have for many decades produced a wealth of aggressive, ambitious young sprinters. Their problem, though, was that for a long time they didn't have adequate college-level training resources for these promising teenagers. Routinely, the very best athletes would leave the country for Britain (Linford Christie) or Canada (Ben Johnson) and often never return.

Then, in the 1970s, former champion sprinter Dennis Johnson did come back to Jamaica to create a college athletic programme based on what he'd experienced in the United States. That programme, now at the University of Technology in Kingston, became the new core of Jamaican elite training. After a critical number of ramp-up years, the medals started to pour in. It was the final piece in the systemic machinery driven by national pride and an ingrained sprinting culture.

Psychology was obviously a critical part of the mix. 'We genuinely believe that we'll conquer,' says Jamaican coach Fitz Coleman. 'It's a mindset. We're small and we're poor, but we believe in ourselves.' On its own, it might seem laughable that self-confidence can turn a tiny island into a breeding ground for champion sprinters. But taken in context of the developmental dynamic, psychology and motivation become vital. Science has demonstrated unequivocally that a person's mind-set has the power to dramatically affect both short-term capabilities and the long-term dynamic of achievement. In Jamaica, sprinting is a part of the national identity. Kids who sprint well are admired and praised; their heroes are sprinters; sprinting well provides economic benefits and ego gratification and is even considered a form of public service.

All things considered, it seems obvious that the mind is the most athletic part of any Jamaican athlete's body.

The notion that the mind is of such paramount importance to athletic success is something that we all have to accept and embrace if we're going to advance the culture of success in human society. Within mere weeks of British runner Roger Bannister becoming the first human being to crack the four-minute mile, several other runners also broke through. Bannister himself later remarked that while biology sets ultimate limits to performance, it is the mind that plainly determines how close individuals come to those absolute limits.

And we keep coming closer and closer to them. 'The past century has witnessed a progressive, indeed remorseless improvement in human athletic performance,' writes South African sports scientist Timothy David Noakes. The record speed for the mile, for example, was cut from 4:36 in 1865 to 3:43 in 1999. The one-hour cycling distance record increased from 26 kilometres in 1876 to 49 kilometres in 2005. The 200-metre freestyle swimming record decreased from 2:31 in 1908 to 1:43 in 2007. Technology and aerodynamics are a part of the story, but the rest of it has to do with training intensity, training methods, and sheer competitiveness and desire. It used to be that 67 kilometres per week was considered an aggressive level of training. Today's serious Kenyan runners, Noakes points out, will cover 230 kilometres per week (at 6,000 feet in altitude).

These are not superhumans with rare super-genes. They are participants in a culture of the extreme, willing to devote more, to ache more, and to risk more in order to do better. Most of us will understandably want nothing to do with that culture of the extreme. But that is our choice.

PART TWO

CULTIVATING GREATNESS

CHAPTER SEVEN

How to Be a Genius (or Merely Great)

The old nature/nurture paradigm suggests that control over our lives is divided between genes (nature) and our own decisions (nurture). In fact, we have far more control over our genes – and far less control over our environment – than we think.

Are [people] conceived with the capacity to play a number of qualitatively different developmental tunes – in other words, to live alternative lives?

– PATRICK BATESON

By now, the reader has realised that this is not really a book about genius in the conventional sense. It is not an instruction manual about how YOU TOO can become JUST LIKE WILLIAM SHAKESPEARE! or a secret decoder to help you ferret out the hidden geniuses among us.

It is, instead, a simple call to all who aspire to achieve – in any arena and on any level. In a world obsessed with discovering innate abilities, the evidence gathered here offers a refreshing turn, away from the notion of fixed, inborn assets and toward the notion of buildable, developing assets. Now we can admire the greatest of the greats – Shakespeare, Einstein, da Vinci, Dante, Mozart, and so on – without getting trapped in an artificial distinction of *us* (innately ordinary) and *them* (innately great). The new science helps us understand how perfectly ordinary human beings grow up to do good, great, and extraordinary things. It exposes the fallacy of giftedness and the tall tales that keep it alive.

> [Setting: Harvard Square, Cambridge, Massachusetts]
> SKYLAR: How did you do that? Even the smartest people I know – and we do have a few at Harvard – have to study a lot. It's hard.
> WILL: Do you play the piano . . . ? Beethoven – he looked at a piano and *saw* music . . . Beethoven, Mozart, they looked at it and it just made sense to them. They saw a piano and they could play. I couldn't paint you a picture, I probably can't hit the ball out of Fenway Park and I can't play the piano –
> SKYLAR: But you can do my O-chem lab in under an hour.
> WILL: When it came to stuff like that, I could always just play.
>
> – From the film *Good Will Hunting*

Beethoven and Mozart would be rolling in the aisles. In truth, their ability to 'see' music came only after years of intensive work – and in Beethoven's case, after horrific abuse. Consider this more reliable description of Beethoven's childhood:

> Neighbours of the Beethovens . . . recall seeing a small boy 'standing in front of the clavier and weeping.' He was so short he had to climb a footstool to reach the keys. If he hesitated, his father beat him. When he was allowed off, it was only to have a violin thrust into his hands, or musical theory drummed into his head. There were few days when he was not flogged, or locked up in the cellar. Johann also deprived him of sleep, waking him at midnight for more hours of practice.
>
> – EDMUND MORRIS, *Beethoven*, 2005

He was four years old. Nearly twenty years later, Beethoven emerged as an extraordinary performer and a promising composer.

But to assert that either he or Mozart could 'always just play' is like saying that a circus clown could always just juggle.

And yet the myth of innate giftedness will live on as long as human beings do. Today, talk of giftedness still pervades our language, even among scientists who know better. It transcends age, class, geography, and religion.

Why? Because we rely on the myth. A belief in inborn gifts and limits is much gentler on the psyche: *The reason you aren't a great opera singer is because you can't be one. That's simply the way you were wired.* Thinking of talent as innate makes our world more manageable, more comfortable. It relieves a person of the burden of expectation. It also relieves us of distressing comparisons. If Tiger Woods is innately great, we can feel casually jealous of his genetic luck while avoiding disappointment in ourselves. If, on the other hand, each one of us truly believed ourselves capable of Tiger-like achievement, the burden of expectation and disappointment could be profound. *Did I blow my chance to be a brilliant tennis player? What would I have to do right now to become a great painter?* In the GxE world, these are not only difficult questions to answer, they can be painful to ask.

Our new developmental paradigm will therefore require not just a new intellectual leap, but also a moral, psychological, and spiritual leap. It begins with a much wider consideration of our true assets and liabilities, which are not just biological but also economic, cultural, nutritional, parental, and ecological. The consideration of what we *inherit* as opposed to what we *choose* will also require a radical revision. According to the old nature/nurture paradigm, biology (nature) is thrust upon us, while we choose our environment (nurture). In the new paradigm, we recognise the folly of these hard-and-fast distinctions.

Heredity, it turns out, is not as straightforward as we have been taught. Parents do pass on unaltered DNA to children, but they also pass on additional instructional material – known as epigenetic material – which helps guide how the genes will be expressed.[4]

[4] A much more thorough explanation of epigenetics can be found in chapter 10.

While genes themselves do not change (by and large) from generation to generation, the epigenetic instructions *can* change. This means that we *can* impact our genetic legacy.

So much for the old black-and-white view of 'nature'.

Meanwhile, we don't really have the control over our environment that we have so long assumed. To begin with a simple example: food. We theoretically choose what we eat, but in reality almost all of us conform to established cultural norms – we eat what our family eats, what our friends and neighbours eat, what our local community eats, what our nation eats. The same principle applies to our language and idioms, to the information and entertainment we consume, our kids' schools and activities, the art and aesthetics we're surrounded by, the people we spend our time with, the basic philosophical notions we subscribe to, and even the air, water, and physical environment that surround us. Even in a land of free choice, we are mostly shaped by habits, messages, schedules, expectations, social infrastructure, and natural surroundings that are not exclusively our own. Many of these elements are passed down from generation to generation with little or no change and are difficult or impossible to alter.

Nothing in this book, therefore, is meant to suggest that any of us have complete control over our lives or abilities – or that we are anything close to a blank slate. Rather, our task now is to replace the simplistic notions of 'giftedness' and 'nature/nurture' with a new landscape: a vast array of influences, many of which are largely out of our control but some of which we can hope to influence as we increase our understanding.

This is a difficult notion and must be allowed to sink in gently. The strong temptation will always be to revert back to the nature/nurture paradigm: if it's not nature, it must be nurture. If it's not genes, it must be environment. If it's not DNA, it must be parenting. But these either/or dichotomies are as misleading as saying that if a person isn't white he must be black. We cannot allow ourselves to think that way anymore.

So, for example, while there's no evidence at all that musical talent sprouts from genes, it does not follow that every person has the necessary resources and tools at any age to accrue prodigious musical skills. There could be any number of limiting factors: inadequate early exposure to music, lacklustre early brain development, inhospitable family and peer attitudes, poor music education, lack of practice time, lack of motivation, mediocre listening habits, lack of suitable mentor, and so on. These are just some of the actual reasons why each five-year-old has a different level of apparent musical 'talent'. Same for every ten-year-old and every thirty-five-year-old. Freedom from genetic oppression doesn't make us all equal, or truly free.

In sum, while our genes may not keep us from greatness, so many other factors can – some of which we unwittingly contribute to, and many of which may be entirely outside of our awareness and/or control.

What about you: Can *you* be a musical genius? A great poet? A world-class chef? It's easy to look at yourself and say, 'Impossible.' But the simple truth is, no one can make such a judgement early in the process. 'The most reasonable assumption seems to be that talent is much more widely distributed than its manifestation would suggest,' wrote talent experts Mihály Csikszentmihályi, Kevin Rathunde, and Samuel Whalen in a 1993 study.

Some guiding principles for the ambitious:

FIND YOUR MOTIVATION.

The single greatest lesson from past ultra-achievers is not how easily things came to them, but how irrepressible and resilient they were. You have to want it, want it so bad you will never give up, so bad that you are ready to sacrifice time, money, sleep, friendships, even your reputation (people may – probably will – come to think of you as odd). You will have to adopt a particular lifestyle of ambition, not just for a few weeks or months but for years and years and years. You have to want it so bad that you are not only ready to fail, but you actually want to experience failure: revel in it, learn from it. It's impossible to say for how long you will have to do these things. You cannot know the results in advance.

Uncommon achievement requires an uncommon level of personal motivation and a massive amount of faith.

The source of motivation is often mysterious, but not always. One of the quirks of human emotion and psychology is that deep motivation can have more than one possible origin. A person can become joyfully inspired, spiritually devoted, or deeply resentful; motivation can be selfish or vengeful, or arise out of a desperation to prove someone right or wrong; it can be conscious or unconscious.

The 1981 movie *Chariots of Fire* highlights the very different motivations of two Olympic runners in the 1920s, Eric Liddell and Harold Abrahams. The devout Christian Liddell runs for the glory of God. 'I believe that God made me for a purpose,' he says, 'but He also made me fast, and when I run, I feel His pleasure.'

Meanwhile, his rival Abrahams, a Jew resentful of anti-Semitic European culture, runs to prove himself to the Christian society and to get revenge. 'So what now? Grin and bear it?' one of Abrahams's friends asks him.

'No, Aubrey. I'm going to take them on. All of them. One by one – and run them off their feet.'

Inspiration may spout after six weeks of life, or sixty years, or never. Where will *yours* come from? A sibling rivalry? A desire to impress your parents or children? An insatiable hunger to be loved? A straightforward fear of failure?

Perhaps you will find it, even more simply, in something you love to do.

Or perhaps you will find it in the anticipation of future regret. Regret turned out to be the final legacy of Lewis Terman's ill-named Genetic Studies of Genius project. In 1995, three Cornell psychologists did an extensive study of Terman's now-elderly participants. They titled their paper 'Failing to Act: Regrets of Terman's Geniuses'. The profound lesson was that, at the end of their lives, Terman's group had exactly the same sorts of regrets as the rest of the elderly population. They wish they had done more: got more education, worked harder, persevered.

That's one Lewis Terman lesson that we can all learn from.

BE YOUR OWN TOUGHEST CRITIC.

Recall the resonant words of Nietzsche: 'All great artists and thinkers [are] great workers, indefatigable not only in inventing, but also in rejecting, sifting, transforming, ordering.' His observation was dead-on, and timeless.

Hollywood movies suggest that genius is a series of *Eureka!* moments, that true greatness flows effortlessly. We live under the great myth of the perfect first draft. While moments of inspiration do exist, great work is, for the most part, painstaking and cannot happen without the most severe (and constructive) self-criticism.[5]

BEWARE THE DARK SIDE (BITTERNESS AND BLAME).

Just as practitioners of judo turn an opponent's attacking energy and momentum into weakness, those with high ambition must constantly turn failure to opportunity. If left to fester into humiliation or bitterness, defeat can take a powerful toll. 'I wake up sometimes and say, "What the heck happened to me?" It's like a nightmare,' American runner Abel Kiviat told the *Los Angeles Times*

[5] It might be fitting to take a moment here and write a few words about how difficult it is for me to get my own writing to the point where I am pleased with it. (Please note I'm making no claims here for what others may think of my work – I'm speaking only about my own opinion.)

It took me nearly three years to write this book. A quick calculation: forty thousand words of text produced over five thousand hours of work comes to, yes, *eight words per hour*. While there are, of course, all sorts of mitigating circumstances, including many hours of research, eight words an hour is actually a pretty good description for how much I accomplish from day to day. My attitude towards my own writing is simple: I assume that everything I write is rubbish until I have demonstrated otherwise. I will routinely write and rewrite a sentence, paragraph, and/or chapter twenty, thirty, forty times – as many times as it takes to feel satisfied. I give myself no time limits. If, on a fresh reading, I am pleased and can't see any way to improve it, I move on. I generally do not begin a subsequent chapter until I am pleased with the current one. In the case of this book, I spent almost a year working on the first chapter alone—and even after that I went back and rewrote it two or three times later on. I do not claim this is the best way to write, only that it works for me.

in 1990 about his disappointing silver medal in the 1,500-metre Olympic run. Kiviat was ninety-one when he made this statement – his performance had occurred more than seventy years earlier!

Unless they somehow fuel motivation, feelings of regret and blame dangerously distract from the task at hand, which is to focus constantly on how to improve.

The worst kind of blame, and the most common, is on one's own biology. This is the great final irony of genetic determinism: the very belief of possessing inferior genes is perhaps our greatest obstacle to success.

IDENTIFY YOUR LIMITATIONS – AND THEN IGNORE THEM.

The pursuit of greatness never makes logical, 'kitchen table' sense. Any possible achievements are years off, far from certain, and often difficult even to envision. The practical distance between your current ability and your desired ability is so enormous that your goal will appear to you and anyone near you as simply unattainable. You are obviously not quick enough, tall enough, strong enough; your intonation isn't true enough; your strokes aren't smooth enough; your material isn't funny enough or sad enough or deep enough; you are *mediocre*. How could you possibly expect to be great?

And that's exactly the point. Greatness isn't just one step beyond mediocrity; it transcends mediocrity, and it does so by taking one step beyond, then another step beyond, then another step beyond – hundreds of thousands of tiny steps until the distance can neither be measured nor even fathomed. The only way to get there is to go farther, harder, longer than almost everyone else, to push well past the point of logic or reason. If it looked easy or even attainable to most, then many more would get there.

That is why ultra-achievers (of whatever age) are also dreamers. They must have part of their heads stuck in the clouds in order to imagine the unimaginable. They have to ignore obvious shortcomings and what may often look like immovable obstacles. To defer to impediments would amount to instant defeat.

In some respects, committing to this pursuit will make even less practical sense as you get older. With every year, you have less time, less schedule flexibility, less energy, and less brain and muscle plasticity. Given the short-term and long-term commitments involved, it is obviously far more possible for an unmarried twenty-year-old to practise deliberately and intensively for hours every day than a married forty-five-year-old with two young kids and a jumbo mortgage. But thousands of extraordinarily successful achievers will attest that there is no age of impossibility. And in some fields, the wisdom that sometimes accompanies age is an asset that cannot be accrued any other way. 'You know, it's interesting,' says a veteran New York magazine and book editor. 'The best writers at age twenty-five are very rarely among the best writers at age fifty. Just staying in the game is difficult, and for those that do, there's a process of quiet, incremental improvement over time that has no substitute. I've found that time is a crucial input into excellence.'

DELAY GRATIFICATION AND RESIST CONTENTEDNESS.

In consumer culture, we are constantly conditioned to gratify our impulses immediately: buy, eat, watch, click – *now*. High achievers transcend these impulses.

Like the Buddha who waits patiently at the gates of heaven until all others have entered before him, young Kenyans are content to run for many years before they can even dream of competing in a major international contest. The tiny violinist screeches out earsplitting sounds not because he thinks a dazzling concerto is right around the corner, but because there is something satisfying in the struggle and in the tiny improvements made along the way. The big prize is envisioned and appreciated as a far-off goal – it is not lusted after. Small accomplishments along the way provide more than enough satisfaction to continue.

HAVE HEROES.

Heroes inspire, not just by their great work but also by their humble beginnings. Einstein worked as a patent clerk. Thomas Edison was expelled from the first grade (age 6–7) because his teacher thought him retarded. Charles Darwin had so little to show for himself as a teenager that his father said to him, 'You care for nothing but shooting, dogs, and rat-catching and you will be a disgrace to yourself and all your family.' (Just a few years later, young Darwin set out on the HMS *Beagle* and eventually revolutionised humanity's view of itself.)

To know the particulars of a favourite artist or athlete's ordeal is to be continually reminded of uncharted paths and oddball ideas that only later become recognised as genius. This experience is magnified by examining rough drafts of masterpiece books, paintings, and albums. To see the evolution of a particular work of art is to behold how *nothing* slowly and painfully becomes *something*. Or, as the legendary musician and artist Brian Eno put it:

> What would be really interesting for people to see is how beautiful things grow out of shit . . . Nobody ever believes [that it happens that way]. Everybody thinks that Beethoven had his string quartets completely in his head, that it somehow appeared there and formed in his head, and all he had to do was write them down . . . What would really be a lesson that everybody should learn is that . . . things come out of nothing. Things evolve out of nothing. The tiniest seed in the right situation turns into the most beautiful forest, and then, the most promising seed in the wrong situation turns into nothing . . .
>
> I think this would be important for people to understand because it gives people confidence in their own lives to know that that's how things work. If you walk around with the idea that there are some people who are so gifted, that they have these wonderful things

> in their head, but you're not one of them, you're just sort of . . . a 'normal' person. [But with this insight], you could have another kind of life. You could say, 'Well, I know that things come from nothing very much and start from unpromising beginnings, and I'm an unpromising beginning – I could start something.'

'Another kind of life'. Here, the artist Eno bumps into the biologist Bateson – who has written of our built-in capacity to 'live alternative lives'. Perhaps there's something to this developmental paradigm after all.

FIND A MENTOR.

Any person lucky enough to have had one great teacher who inspired, advised, critiqued, and had endless faith in her student's ability will tell you what a difference that person has made in her life. 'Most students who become interested in an academic subject do so because they have met a teacher who was able to pique their interest,' write Csikszentmihályi, Rathunde and Whalen. It is yet another great irony of the giftedness myth: in the final analysis, the true road to success lies not in a person's molecular structure, but in his developing the most productive attitudes and identifying magnificent external resources.

CHAPTER EIGHT

How to Ruin (or Inspire) a Kid

Parenting does matter. There is much parents can do to encourage their kids to become achievers, and there are some important mistakes to avoid.

Do we know how many geniuses are never recognised because their talents are blighted before they have a chance to be expressed? The fact is, nobody does.

– TALENT RESEARCHERS MIHALY CSIKSZENTMIHALYI, KEVIN RATHUNDE AND SAMUEL WHALEN

To say that there is much we don't control in our lives is a dramatic understatement, roughly on the order of saying that the universe is a somewhat large place. To begin with, there are many influences we can't even detect. In 1999, Oregon neuroscientist John C. Crabbe led a study on how mice reacted to alcohol and cocaine. Crabbe was already an expert on the subject and had run many similar studies, but this one had a special twist: he conducted the exact same study at the same time in three different locations (Portland, Oregon; Albany, New York; and Edmonton, Alberta) in order to gauge the reliability of the results. The researchers went to 'extraordinary lengths' to standardise equipment, methods, and lab environment: identical genetic mouse strains, identical food, identical bedding, identical cages, identical light schedule, etc. They did virtually everything they could think of to make the environments of the mice the same in all three labs.

Somehow, though, invisible influences intervened. With the scientists controlling for nearly everything they could control, mice with the exact same genes behaved differently depending on where they lived. And even more surprising: the differences were not consistent, but zigged and zagged across different genetic strains and different locations. In Portland, one strain was especially sensitive to cocaine and one especially insensitive, compared to the same strains in other cities. In Albany, one particular strain – just the one – was especially lazy. In Edmonton, the genetically altered mice tended to be just as active as the wild mice, whereas they were more active than the wild mice in Portland and less active than the wild mice in Albany. It was a major hodgepodge.

There were also predictable results. Crabbe did see many expected similarities across each genetic strain and consistent differences between the strains. These were, after all, perfect genetic copies being raised in painstakingly identical environments. But it was the unpredicted differences that caught everyone's attention. 'Despite our efforts to equate laboratory environments, significant and, in some cases, large effects of site were found for nearly all variables,' Crabbe concluded. 'Furthermore, the pattern of strain differences varied substantially among the sites for several tests.'

Wow. This was unforeseen, and it turned heads. Modern science is built on standardisation; new experiments change one tiny variable from a previous study or a control group, and any changes in outcome point crisply to cause and effect. The notion of hidden, undetectable differences throws all of that into disarray. How many assumptions of environmental sameness have been built right into conclusions over the decades? What if there really is no such thing?

What if the environment turns out to be less like a snowball that one can examine all around and more like the tip of an iceberg with lurking unknowables? How does that alter the way we think about biological causes and effects?

Something else stood out in Crabbe's three-city experiment: gene-environment interplay. It wasn't just that hidden environmental differences had significantly affected the results. It was also clear that these hidden environments had affected different mouse strains in different ways – clear evidence of genes interacting dynamically with environmental forces.

But the biggest lesson of all was how much complexity emerged from such a simple model. These were genetically pure mice in standard lab cages. Only a handful of known variables existed between groups. Imagine the implications for vastly more complex animals – animals with highly developed reasoning capability, complex syntax, elaborate tools, living in vastly intricate and starkly distinct cultures and jumbled genetically into billions of unique identities. You'd have a degree of GxE volatility that would boggle any scientific mind – a world where, from the very first hours of life, young ones experienced so many hidden and unpredictable influences from genes, environment, and culture that there'd be simply no telling what they would turn out like.

Such is *our* world. Each human child is his/her own unique genetic entity conceived in his/her own distinctive environment, immediately spinning out his/her own unique interactions and behaviours. Who among these children born today will become great pianists, novelists, botanists, or marathoners? Who will live a life of utter mediocrity? Who will struggle to get by? We do not know.

What we do know is that our brains and bodies are primed for plasticity; they were built for challenge and adaptation. This is true from life's earliest moments. According to neuroscientists Mark H. Johnson and Annette Karmiloff-Smith, 'Recent reviews of pre- and postnatal brain development have come to the conclusion that brain development is not merely a process of the unfolding of a genetic plan, or a passive response to the environmental input, but is an activity-dependent process at the molecular, cellular, and organismal levels involving probabilistic epigenesis (bidirectional relations between genes, brain and behaviour).'

Put more simply: 'Human babies are special,' says Andrew Meltzoff, codirector of the Institute for Learning and Brain Sciences at the University of Washington. 'What makes them special is not that they are born so intelligent but that they are designed to change their minds when faced with the data.'

Intelligence is not fixed but waiting to be developed. Athletic prowess is not preordained but awaits training. Musical ability lies dormant in all of us, calling for early and sustained incantation. The potential for creativity is built into the architecture of our brains. All of these are a function of influence and process – far from fully controllable, but also quite the opposite of fixed and predetermined.

The parent's job, then, is to respect and engage in that process – which has, of course, already started long before birth. Every parent experiences the odd sensation of getting to know his or her newborn child, recognising in him or her a unique personality that seems to have been already formed. That's because the process is already nine months old. The process has already begun.

On reflection, we parents are not so far off from John Crabbe and his mice. In his lab, Dr. Crabbe studies the interaction between environment and mouse genome. In our home nursery, we also watch how our child's unique biology interacts with various facets of the outside world: we see what makes her laugh and cry, what grabs her attention and bores her silly, what tastes good and bad. We get to know not her preset design, but how she responds to different versions of the world we present to her.

Based on our reading of these interactions, we then tailor her environment accordingly. We mesh our own aspirations with what we learn about the child.

This is the ultimate lesson of GxE: rather than first waiting for our natural gifts to sprout, we must immediately dive into the process, embracing the inseparability of nature and nurture. We know that genes are playing a key role and that their expression is being determined every moment by the quality of the life our child leads. We know that we are helping to choose our child's own jukebox tune. Our job is to find the process that produces the best possible individual.

Of course, one does not have to aim for a gold medal to incorporate the lessons of talent and ability that come from this book. There are many quietly heroic ways to be a modest or terrific success: a wonderful teacher, a sharp and creative and ethical entrepreneur, even a loyal and hardworking assistant or clerical worker.

Ultimately, of course, the life goal will be up to the individual. But parents can sow certain seeds and water them.

Or can they? In 1998, writer Judith Rich Harris shook the world of academic psychology with her book *The Nurture Assumption*. 'Do parents have any important long-term effects on the development of their child's personality?' she asked, then bluntly declaring, 'The answer is no.' Relying heavily on the identical-twin heritability studies from the 1980s and '90s (discussed in chapter 4), Harris concluded that parents are more genetic guardians of their children's personalities than they are active shapers of them. The most important environmental influences on character, she proposed, are not parents but peers.

Challenging assumptions is always healthy, and in one sense Harris's book was a welcome critique that forced university psychologists out of their comfort zone. But a decade later, her argument is a victim of its own stale assumptions, beginning with her stance on genetics. 'Genes contain the instructions for producing a physical body and a physical brain,' Harris wrote. 'They determine the shape of the facial features and the structure and chemistry of the brain. These physical consequences of heredity are the straightforward consequences of carrying out the instructions in the genes; I call them *direct genetic effects*.'

This was an understandable view in 1998, but now we know better. We know now that there are no real 'direct genetic effects' and that the nature/nurture distinction is a false one.

Saddled by the old view of genetics, Harris believed that 50 per cent of a person's character comes straight from his or her genes, while most of the rest comes from what behavioural psychologists were calling 'non-shared' environment – a term proposed by geneticist Robert Plomin to explain not-yet-understood environmental influences. The ambiguous word 'non-shared' was designed to convey the opposite of shared family experiences that researchers had assumed affected siblings in similar ways. Non-shared experiences, they reasoned, would affect siblings differently. Much of Harris's book is an effort to convince the world that peers are the crucial non-shared influence in kids' lives.

Two years after her book came out, though, it turned out that there was a problem with the shared/non-shared paradigm. An analysis in 2000 by the University of Virginia psychologist and expert in behavioural genetics Eric Turkheimer revealed that it was another false distinction. Just like 'nature/nurture' was supposed to separate genetic effects from environmental effects, 'shared' and 'non-shared' implied that it was either/or: either people would have similar reactions to shared experiences *or* they would have different reactions to non-shared experiences. Turkheimer's powerful meta-analysis revealed the much more common third possibility: most of the time, kids have different reactions to shared experiences. (As Turkheimer put it more clinically: 'Non-shared environmental variability predominates not because of the systematic effects of environmental events that are not shared among siblings, but rather because of the unsystematic effects of *all* environmental events.')

Harvard psychologist Howard Gardner had an even more fundamental problem with Harris's notion of uninfluential parents. 'When we consider the empirical part of Harris's argument,' he wrote in the *New York Review of Books*, 'we find it is indeed true that the research on parent-child socialisation is not what we would hope for. However, this says less about parents and children and more about the state of psychological research, particularly with reference to "softer variables" such as affection and ambition. While psychologists have made genuine progress in the study of visual perception and measurable progress in the study of cognition, we do not really know what to look for or how to measure human personality traits, individual emotions, and motivations, let alone character.'

'My reading of the research,' Gardner continued, 'suggests that, on the average, parents and peers will turn out to have complementary roles: parents are more important when it comes to education, discipline, responsibility, orderliness, charitableness, and ways of interacting with authority figures. Peers are more important for learning cooperation, for finding the road to popularity, for inventing styles of interaction among people of the same age. Youngsters may find their peers more interesting, but they will look to their parents when contemplating their own futures . . . I would give much weight to the hundreds of studies pointing towards parental influence and to the folk wisdom accumulated by hundreds of societies over thousands of years. And I would, accordingly, be sceptical of a perspective, such as Ms. Harris's, that relies too heavily on heritability statistics and manages to reanalyse numerous studies and practices so that they all somehow point to the peer group.'

So yes, parents matter. Parenting isn't everything or the only thing. Parents don't have anything close to complete control and in most cases should not shoulder all the blame when things don't turn out well. But parenting does matter. And to the extent that parents can have a serious impact on the goals, strategies, and personal philosophies of their children, here are four key guideposts to excellence:

1. BELIEVE

In 1931, a young Japanese violinist and instructor named Shinichi Suzuki was teaching a violin class composed mostly of young men. After class one day he was approached by the father of a four-year-old boy: would he consider teaching the gentleman's son?

Suzuki was startled and dumbfounded. He had no idea if a four-year-old could learn to play the violin and little idea how to instruct him. While rehearsing shortly afterwards, though, a profound thought struck him: virtually all Japanese children learn to speak Japanese – early, and with precision. 'The children of Osaka speak the difficult Osaka dialect,' Suzuki thought to himself. '[They] are unable to speak the Tohoku dialect, but the Tohoku children speak it. Isn't that something of an accomplishment?'

The obvious lesson, Suzuki surmised, was this: through extraordinary repetition, parental persistence, and strong cultural reinforcement, every young child masters this steep technical challenge. Why couldn't this lesson apply just as directly to music?

So Suzuki did accept four-year-old Toshiya Eto as a pupil and began to develop a method of instruction he called the 'mother-tongue method'. He emphasised heavy parent involvement, steady practice, memorisation, and lots of patience. (In retrospect, the parallels between Suzuki's approach and young Mozart's musical development are uncanny.) Little Toshiya Eto responded beautifully, prompting Suzuki to recruit more young pupils and refine his methods further. He came to quickly believe, in fact, that early musical training has an overwhelming advantage over later training and that it was a gateway to an enlightened life.

He also began to attract attention. A few years into his radical experiment, Suzuki featured seven-year-old Toshiya and several other young students in a public performance. A local newspaper became fixated on the marvels of three-year-old Koji Toyoda, who played one of Dvořák's 'Humoresques' on a one-sixteenth-size violin. 'A Genius Appears!' ran the headline. Suzuki was horrified by this interpretation. '[Before the concert], I had told journalists: talent is not inherent or inborn, but trained and educated . . . I had put emphasis on this and had repeated it.' The message was just as important to Suzuki as his method: gifts and talents, he was convinced, were not exclusive to the privileged few; with the right training and persistence, anyone could achieve remarkable success.

As his first young pupil, Toshiya Eto, developed into a world-renowned musician, Suzuki continued to refine his methods and spread their application. By 1949, his Talent Education Research Institute had thirty-five branches in Japan and was teaching fifteen hundred children. The Suzuki method became a sensation around the world and helped transform our understanding of young children's capabilities.

It begins with a simple faith that each child has enormous potential and that it is up to us to muster whatever resources we can to exploit that potential. Rather than wonder if their child is among the 'gifted' chosen few, parents should believe deeply in the extraordinary potential of their children. Without that parental faith, it is highly unlikely that significant achievement will occur.

2. SUPPORT, DON'T SMOTHER

Imagine, for a moment, that the day your child is born, the doctor gives you a choice of two infant nutritional supplements. The first will transform your child into an astonishing prodigy who, in adulthood, will probably fall back into mediocrity and possibly develop severe emotional problems. The second will produce an emotionally balanced child who is highly unlikely to be a tiny star athlete or musician early on, but who will slowly gather the tools to become a confident, enlightened person with solid relationships and a deep belief in the value of hard work. In the long run, he will have the resources to achieve greatness as an adult.

This stark choice may seem a little absurd, but unconsciously, it is the choice that many parents make.

'You could call it the Britney Spears Syndrome,' says Columbia University psychiatrist Peter Freed. 'I see it frequently in my practice – a clear model for how the narcissistic parent injures a child's sense of self by attaching high-achievement to love.'

It all begins, explains Freed, with a parent who has grown up believing that, in order to be liked, he must be exceptional in some way. The parent subsequently showers his own children with affection after each accomplishment and shuns them after failure. 'The parent beams when the child performs well, and then withdraws love when he's underperforming,' says Freed. 'The kid becomes addicted to pleasing the parent. When he doesn't live up to the parent's expectations, he feels his parent go cold, which of course is totally devastating. That on-again off-again feeling about how love works sets the stage for narcissism.' In early adulthood, Freed explains, the child will inevitably struggle with social and emotional challenges (as everyone does) and find that he doesn't have a deep emotional reservoir to fall back on. The foundations of love and trust are corrupted by what he experienced as a child. The child victim of a narcissistic parent frequently has a difficult time forming stable life partnerships.

The flip side, says Freed, is a parent who offers unconditional and unshifting love that is decidedly not connected to achievement.

'Non-narcissistic parents follow the child's lead,' he explains. 'They're very good at limit-setting and setting high expectations, but they will wait to see what the kid wants to do and not become anxious if he isn't high-achieving early on. Their attitude is that the most important thing you're doing in childhood is making friends and being an active part of the community. If the team wins, they'll be happy, but if the team has trouble, they'll have them over and watch a movie.'

There is, in other words, a right way and a wrong way to direct your kids toward achievement. Early exposure to resources is wonderful, as is setting high expectations and demonstrating persistence and resilience when it comes to life challenges. But a parent must not use affection as a reward for success or a punishment for failure. The parent must show faith in the child's ability to seek achievement for his or her own inner satisfaction.

3. PACE AND PERSIST

'It's not that I'm so smart,' Albert Einstein once said. 'It's just that I stay with problems longer.'

Einstein's simple statement is a clarion call for all who seek greatness, for themselves or their children. In the end, persistence is *the* difference between mediocrity and enormous success.

The big question is, can it be taught? Can persistence be nurtured by parents and mentors?

Boston College's Ellen Winner insists not. Persistence, she argues, 'must have an inborn, biological component.' But the evidence indicates otherwise. The brain circuits that modulate a person's level of persistence are plastic – they *can* be altered. 'The key is intermittent reinforcement,' says Robert Cloninger, a Washington University biologist. 'A person who grows up getting too frequent rewards will not have persistence, because they'll quit when the rewards disappear.'

This jibes well with Anders Ericsson's finding about deliberate practice and with the ascetic philosophy of Kenyan runners: an emphasis on instant gratification makes for bad habits and no effective long-term plan. The ability to delay gratification opens up a whole new vista for anyone looking to better herself.

It also conjures up a classic study by Stanford psychologist Walter Mischel, who in the early 1970s offered a group of four-year-olds a choice: they could have one marshmallow immediately or wait a short while (until the researcher got back from an 'errand') for two marshmallows. The results:

- One-third of the kids immediately took the single marshmallow.
- One-third waited a few minutes but then gave in and settled for the single marshmallow.
- One-third patiently waited fifteen minutes for two marshmallows.

At the time, it impressed Mischel and his colleagues that so many very young children had the self-discipline to wait indefinitely for a larger reward. But the real lesson came after fourteen years of Mischel's own waiting – until his original subjects had taken the SATs and were finishing high school. Comparing the SAT scores of the original nonwaiting (instant gratification) group to the waiting (delayed gratification) group, he found the latter scored an average of 210 points higher. Those with an early capacity for self-discipline and delayed gratification had gone on to much higher academic success. The delayed-gratification kids were also rated as much better able to cope with social and personal problems.

The marshmallow study also demonstrated the ability to develop such skills. In side experiments, researchers transformed kids' wait times by suggesting how to think of the rewards. When kids staring at real marshmallows were encouraged to imagine them as pictures of marshmallows – making them more abstract in their minds – it lengthened their ability to wait from six to eighteen minutes. (The reverse was also true – kids imagining pictures as real marshmallows had their waiting ability shortened.)

Strategies like these prove that a kid's mode of gratification can be altered by parents and teachers. Overall, what emerges about the study of delayed gratification is that it is a skill set – and the skills can be acquired. Kids can learn to distract themselves from objects of desire, learn to abstract those desires, learn to monitor their own progress, and so on. 'Children will have a distinct advantage beginning early in life,' Mischel concluded, 'if they use effective self-regulatory strategies to reduce frustration in situations in which self-imposed delay is required to attain desired goals.'

Any parent can adopt basic strategies to encourage self-discipline and delayed gratification. Here are two:

- **Model self-control.** Behave as you'd want your child to behave, now and in the future. Don't buy, eat, or grab whatever you want whenever you want it. The more self-control you demonstrate, the more your child will absorb.
- **Give kids practice.** Don't immediately respond to their every plea. Let them learn to deal with frustration and want. Let them learn how to soothe themselves and discover that things will be all right if they wait for what they want.

There's no single pathway to achieve these desired results as a parent, of course. Each parent must chart his or her own course. Any philosophy, religion, or practical exercise that reinforces that principle is going to work well for parents and children.

4. EMBRACE FAILURE

In the sometimes counterintuitive world of success and achievement, weaknesses are opportunities; failures are wide-open doors. The only true failure is to give up or sell your children short.

Developmental biologists, in fact, stress that all of human development is set up to be a response to problems and failures. Parents are supposed to play an important role by drawing attention to those challenges. 'Specific motor problems are in many cases called to the infant's attention or even thrust upon the infant by one or more caretakers in what we call a field of promoted action,' writes the noted philosopher of science Edward S. Reed and his colleague Blandine Bril. 'It is because human adults promote specific motor problems for infants – often before the child is capable of solving that problem – that human action development takes the course that it does.'

In other words, parents are not supposed to make things easier for kids. Instead, they are supposed to present, monitor, and modulate challenges. The great success stories in our world come about when parents and their children learn to turn straight into the wind and gain satisfaction from marching against its ever-increasing force.

CHAPTER NINE

How to Foster a Culture of Excellence

It must not be left to genes and parents to foster greatness; spurring individual achievement is also the duty of society. Every culture must strive to foster values that bring out the best in its people.

That whole philosophy of persistence . . . is one that I'm going to be emphasising again and again in the months and years to come, as long as I am in this office. I'm a big believer in persistence. I think that . . . if we keep on working at it, if we acknowledge that we make mistakes sometimes and that we don't always have the right answer, and we're inheriting very knotty problems, that we can pass health care, we can find better solutions to our energy challenges, we can teach our children more effectively . . . I'm sure there'll be more criticism and we'll have to make more adjustments, but we're moving in the right direction.

– PRESIDENT BARACK OBAMA, March 24, 2009

Leonardo da Vinci, painter of *Mona Lisa* and *The Last Supper*, exceptional engineer and anatomist, conceptualist of the automobile, helicopter, and machine gun, and also part-time geographer, mathematician, musician, and botanist, considered by some historians to be the most diversely talented person in the history of humankind, could also be a bit of a jerk. According to the sixteenth-century artist and writer Giorgio Vasari (a direct witness), da Vinci sported a public 'disdain' for his younger peer Michelangelo Buonarroti – a hostility so strong that the great Michelangelo eventually felt compelled to leave Florence so that he and Leonardo wouldn't have to share the same town. Da Vinci also pointedly criticised the art of sculpture – Michelangelo's forte – as a messy, easier, and obviously inferior craft that requires 'greater physical effort [while] the painter conducts his works with greater mental effort.'

Not that Michelangelo treated his elder rival any better. His general disposition toward Leonardo was said to be resentful and mean-spirited. On one occasion when the two men happened to be in the same vicinity, a bystander's comment led to a rather nasty exchange:

> Walking with a friend near S. Trinità, where a company of honest folk were gathered, and talk was going on about some passage from Dante, they called to Lionardo, and begged him to explain its meaning. It so happened that just at this moment Michelangelo went by, and, being hailed by one of them, Lionardo answered: 'There goes Michelangelo; he will interpret the verses you require.' Whereupon Michelangelo, who thought he spoke this way to make fun of him, replied in anger: 'Explain them yourself, you who made the model of a horse to cast in bronze, and could not cast it, and to your shame left it in the lurch.' With these words, he turned his back to the group, and went his way. Lionardo remained standing there, red in the face for the reproach cast at him; and Michelangelo, not satisfied, but wanting to sting him to the quick, added: 'And those Milanese capons believed in your ability to do it!'

Today, we gaze at the *Mona Lisa* and the statue of David as phenomenal works rendered by singular geniuses, and we pay little mind to the gritty human process behind their creation. In so doing, though, we often overlook what may be the central cultural lesson of great achievement: that it is rooted in comparison and rivalry. 'Every natural gift must develop itself by contests,' wrote Nietzsche. While we tend of think of achievement as an individual phenomenon, no human is an island. At its essence, humanity is a social and competitive enterprise. We learn from one another, share with one another, and constantly compare and compete with one another for affection, accomplishment, and resources.

It cannot, then, simply be left to genes, vitamins, and parents to foster greatness; spurring individual achievement also must be the duty of society. Every culture must strive to foster values that bring out the best in its people.

Cultural differences matter enormously. In the seventh and eighth centuries, the Islamic Renaissance radiating from Baghdad sparked great advances in agriculture, economics, law, and literature. Mathematicians used spherical trigonometry and the new science of algebra to develop a more precise calculation of time, latitude and longitude, the earth's surface area and circumference, and the location of the stars. Europe at the time had nothing like this same inventiveness; it would have to wait until the twelfth century for its analogous culture of innovation. (Among other developments, there were twelfth-century European advances in printing, timekeeping, astronomy, navigation, lenses, ships and guns.)

History is filled with hundreds of such achievement clusters and achievement black holes.

In the eighteenth and nineteenth centuries, France revolutionised Western cooking with dramatic new sauces, soufflés, soups, and pastries, while nearby England rested with its sweet and savoury meat pies. In the twenty-first century, the United States is home to eleven of the fifteen top-rated universities in the world; the entire African continent doesn't have even one university in the top 150.

Around 1900, the single city of Vienna incubated the work of Gustav Klimt, Gustav Mahler, Arnold Schoenberg, Otto Wagner, Sigmund Freud, and Ludwig Wittgenstein. In the 1980s and '90s, the modest region known as Silicon Valley, just south of San

Francisco, turned out so many innovations in computer hardware and software that it rapidly transformed the very character of human society. Cultural clusters of innovation and excellence can be as regional as New Orleans jazz, as period-specific as mid-twentieth-century Eastern European physics, and as vital to the betterment of humankind as New Haven pizza.

How do some cultures motivate superb achievement while others leave potential geniuses uninspired and inert? In his study of the ancient Greeks, Nietzsche imagined Plato declaring, 'Only the contest made me a poet, a sophist, an orator!' Competition, Nietzsche observed, was central to that culture, where rivalries were encouraged not only in sports but also in oratory, drama, music, and politics. Other Greek historians concur. 'The ancient Greeks turned competition into an institution on which they based the education of their citizens,' explains Olympic official Cleanthis Palaeologos. 'They presented the victory at major games as a godsent blessing, a joy and pride for the city, its fame and prestige, and they recognised the victors as men worthy of respect and honoured them with great distinctions.'

The ambitious goal was to assist as many Greek citizens as possible (though not women or slaves) in their aim to attain the human ideal. To achieve this, public spaces and customs were designed to encourage public education, mentorship, achievement, and the competitive spirit known as 'agonism'. The key emphasis was on contest as a means, not an end. 'Agonism implies a deep respect and concern for the other,' explains political theorist Samuel Chambers. 'Indeed, the Greek *agon* refers most directly to an athletic contest oriented not merely towards victory or defeat, but emphasising the importance of the struggle itself . . . marked not merely by conflict but just as importantly, by mutual admiration.'

With this ideal, the Greeks planted a seed that has sprouted from time to time in cultures enlightened enough to understand its promise. Dutch historian Johan Huizinga suggests that without the agonistic spirit, human beings would simply be incapable of rising above mediocrity.

Which brings us back to the Italian Renaissance, one of the

most concentrated periods of creativity in history. Not coincidentally, it was also an era of planned cultural combat in which patrons and artists constantly competed against one another for the best ideas and works. Leonardo, Michelangelo, Raphael, Titian and Correggio were all open-eyed adversaries who learned from, mimicked, advised, critiqued, annoyed, one-upped, and desperately admired one another. Aesthetic rivalries also flourished on a political level. Interspersed between actual life-and-death battles, cities fought artistic wars, competing against one another for the finest public monuments. As soon as Florence began to build a new colossal *duomo*, for example, Siena immediately set out to exceed it.

In fact, the Italian Renaissance actually began with a specific contest, according to Rutgers art historian Rona Goffen. In the year 1400, Florence's Merchants Guild launched a competition to create grand new doors for its octagonal baptistry. The contest winner, Lorenzo Ghiberti, later reported that seven *combattitori* had competed for the commission and that 'to me was conceded the palm of victory.' After that, such contests gradually became commonplace, and the increasingly competitive arts culture fuelled both public interest and artistic achievement. Artists were pitted against one another like gladiators; bruised feelings were as much a part of the scene as religious inspiration and bold new ideas. In 1503, Piero Soderini, the newly elected chief executive of the Republic of Florence, commissioned Leonardo and Michelangelo to work literally side by side on the walls of the council hall. Da Vinci was asked to depict the battle of Anghiari, Michelangelo the battle of Cascina. The rivalry was exploited to the fullest: the contract specified that they were to be 'in competition with each other'. The public was expected to enjoy the spectacle. 'Artists have always borrowed from each other,' writes Goffen. 'What is different about the sixteenth century is that the great masters . . . often knew each other's major patrons; and they knew each other, sometimes as friends and colleagues, sometimes as enemies – but always as rivals.'

And yes, this rivalry even extended to the great Sistine Chapel. Today, one can stand beneath the majesty of Michelangelo's ceiling frescos in the chapel and take in the full sweep of their glory. At the time of its inception, though, Michelangelo was convinced that his commission from Pope Julius II – which he tried to refuse but could not – was a dangerous sidetrack to his career plotted by the politically savvy Raphael, a much more experienced painter. (Leonardo, meanwhile, was not even invited to compete for the prestigious assignment, which provoked a different sort of resentment.)

The lesson is clear: when we celebrate a great achievement, we are not just celebrating hard work, but also a competitive process where some have won and others have lost. This would be a brutal feature of humanity if we didn't also know – from chapter 3 – that given the right mind-set, failure is good for us.

The problem is that different people have very different attitudes toward competition. In 1938, Harvard psychologist Henry A. Murray proposed that human beings could be separated into two distinct competitive personalities: HAMs ('high in achievement motivation') and LAMs ('low in achievement motivation'). HAMs enjoy and perform better under directly competitive conditions than they do under noncompetitive conditions. LAMs dislike competition, do not seek it out, and are less happy and productive when pushed into it. They do better when pursuing so-called mastery goals – improvement of a skill in comparison to oneself rather than to others.

In Western societies, a higher proportion of men are HAMs and a higher proportion of women are LAMs. Interestingly, though, it turns out that this gender divide is not universal or genetically hardwired. In 2006, economists Uri Gneezy, Kenneth L. Leonard, and John A. List compared competitive instincts in two very different societies: Maasai in Tanzania and Khasi in India. Among the patriarchal Maasai, men choose to compete at twice the rate of women. But among the Khasi, which is rooted in a matrilineal culture where women inherit property and children are named from the mother's side of the family, women choose to compete much more often than men.

The first point to take away from this study is that there is clearly no fixed male or female competitive biology. How men and women act is dependent on cultural circumstances and gene-environment interaction. 'Our results have import within the policy community,' Gneezy and colleagues concluded. 'If the difference is based on nurture, or an *interaction between nature and nurture* . . . public policy might [best] be targeting the socialisation and education at early ages as well as later in life to eliminate this asymmetric treatment of men and women.'

The much larger point is that a person's internal motivation is highly malleable and is closely tied to social reality. Our cultural landscape directly affects whether and how people challenge themselves and others to achieve.

The trick, then, is to sculpt a culture that encourages healthy achievement and that can accommodate different personality types and levels of motivation. How can we best create classrooms, offices, and communities where competitive instincts are rewarded but where less competitive individuals also feel energised rather than suffocated?

Not surprisingly, the answer turns out to be making sure that near-term tasks are clear and meaningful. If short-term tasks can be made relevant to long-term goals, researchers have found, then even LAMs will dive in and relish the challenge. This fits perfectly with Ericsson's 'deliberate practice' – the satisfaction of working hard to master near-term goals, learning to enjoy the process rather than focus on the large gulf between current abilities and the far-off ideal.

It also points clearly to a new direction for schools, which must recognise that abilities are achievable skills and not innate entities (à la Carol Dweck, chapter 5) and must find a way to motivate every child.

Sound too ambitious? Toronto writer and educator John Mighton might have agreed before he became a maths tutor in his late twenties. But after a short time working with so-called learning-impaired students, Mighton was shocked to learn how far and fast they could progress with the right teaching methods. He realised that countless maths students get left behind at one point or another simply because they can't quite grasp one small concept; they then quickly lose confidence in their ability to go forward, and their abilities stagnate. Mighton's response to this problem was to break down mathematical concepts into the most easily digestible form and help students build skills and confidence in tandem. He called his new programme 'Junior Undiscovered Math Prodigies', or JUMP. 'With proper teaching and minimal tutorial support,' he writes in his book *The Myth of Ability*, 'a Grade 3 class could easily reach a Grade 6 or 7 level in all areas of the mathematics curriculum without a single student being left behind. Imagine how far children might go (and how much they might enjoy learning) if they were offered this kind of support throughout their school years.'

Mighton does not claim his particular teaching method as the only approach, or even the best. But 'whatever method is used,' he insists, 'the teacher should never assume that a student who initially fails to understand an explanation is therefore incapable of progressing.'

We know – thanks to Carol Dweck, Robert Sternberg, James Flynn and others – that Mighton is absolutely correct. In fact, countless students fall behind in maths and other subjects for exactly the same reason others generally hate to compete directly in any field: it makes them feel that their permanent limitations are being exposed. People stop striving in a certain area when they receive the message that they simply don't have what it takes. 'I wasn't quite suited for the educational system,' Bruce Springsteen has said of his early days. 'One problem with the way the educational system is set up is that it only recognises a certain type of intelligence, and it's incredibly restrictive – very, very restrictive. There's so many types of intelligence, and people who would be at their best outside of that structure [get lost].'

Schools can adapt to the reality that different people have different ways of learning. It is not a contradiction to maintain high expectations of every student *and* to show compassion and creativity for those who, inevitably, do not immediately meet those expectations. Failure should be seen as a learning opportunity rather than a revelation of students' innate limits. 'If non-linear leaps in intelligence and ability are possible,' writes John Mighton, 'why haven't these effects been observed in our schools? I believe the answer lies in the profound inertia of human thought: when an entire society believes something is impossible, it suppresses, by its very way of life, the evidence that would contradict that belief.'

Set high expectations, but also show compassion, creativity, and patience. This same set of principles applies to other sectors of society and culture. It's how the government should treat its poorest citizens and how the legal system should treat its transgressors. It's how bosses should treat employees and how businesses should treat consumers. It's how the media should treat its audience.

There is a much uglier alternative. We can instead embrace a rawer, purer competitive atmosphere – a winner-take-all system. 'Man – every man – is an end in himself, not the means to the ends of others,' Ayn Rand wrote in 1962. 'He must exist for his own sake . . . The pursuit of his own rational self-interest and of his own happiness is the highest moral purpose of his life.' This is the laissez-faire ideal, the belief that pure self-interest and market efficiencies will create the most productive society.

A laissez-faire society *will* bring great achievement. The most competitive will rise to the top, at the expense of others. Competition will know no moral boundary. Society will, in every way, become more and more extreme, producing some great achievers and many unfortunate losers. Recall *Sports Illustrated*'s Alexander Wolff's analysis of the Kenyan running culture: with a million Kenyan schoolboys running so enthusiastically, Kenyan coaches can afford to push their athletes to the most extreme boundaries, knowing that they will lose many to exhaustion and injury, but that enough will thrive to make their teams successful.

But this sacrificial ethos is not the sort of humanity we seek. Instead, we embrace the agonistic ideal: healthy rivalry, high expectations, respect and compassion for all.

The genius in all of us is that we can all rise together.

CHAPTER TEN

Genes 2.1 – How to Improve Your Genes

We have long understood that lifestyle cannot alter heredity. But it turns out that it can . . .

Over the last century, few scientists' names have been subjected to as much historical derision as early-nineteenth-century French biologist Jean-Baptiste de Lamarck. In textbooks and elsewhere, Lamarckism has been defined (and mocked) as a crude, pre-Darwinian conception of evolution, tainted by the flimsy idea that biological heredity can somehow be altered through personal experience.

Lamarck called it 'the inheritance of acquired characteristics' – the notion that an individual's actions can alter the biological inheritance passed on to his or her children. For example, giraffes, according to Lamarck's theory, had developed longer and longer necks over the generations because of the giraffe's practice of reaching higher and higher for food.

> The giraffe is . . . obliged to browse on the leaves of trees and to make constant efforts to reach them. From this habit long maintained in all its race, it has resulted that the animal's forelegs have become longer than its hind-legs, and that its neck is lengthened.
>
> – JEAN-BAPTISTE DE LAMARCK, *Philosophie Zoologique*, 1809

This sounds preposterous to us now, mostly because it is so different from our Darwinian understanding of evolution. After Darwin's *Origin of Species* and others' subsequent discovery of genes, a very different notion – the theory of natural selection – became scientific and popular consensus. For more than a century, it has been universally accepted that genes are altered not by individual experience but by random mutation and other factors. The individuals whose mutations happen to best fit their environments will thrive and will pass their genes on to future generations.

We cannot change our genes. In the 1950s, the discovery of DNA reaffirmed this idea and secured Lamarck's place in history as the intellectual loser. Today, any high school student knows that genes are passed on unchanged from parent to child, and to the next generation and the next. Lifestyle *cannot* alter heredity.

Except now it turns out that it can . . .

. . .

In 1999, botanist Enrico Coen and his colleagues at the United Kingdom's John Innes Centre were trying to isolate the genetic differences between two distinct types of the toadflax plant.

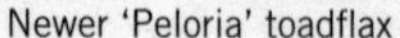

Newer 'Peloria' toadflax

EMIL NILSSON

Ordinary toadflax

The newer and rarer type, named 'Peloria' (left, above) by Carl Linnaeus in the mid-eighteenth century, has a distinct type of flower with five spurs surrounding it like a star.

The trouble was, this difference couldn't be found on the genes. When they looked closely at the gene known to be associated with flower symmetry, a gene known as *Lcyc*, Coen's team was astounded to find that the DNA code in each plant was exactly the same. Two very distinct flowers, same genetic code.

What they discovered next was even more surprising. There *was* a difference between the two flowers on their respective *epigenomes* – the packaging that surrounds DNA.

Some quick background on genetic architecture: DNA is famously wound together in a double-helix strand that, close-up (at a magnification of about 10 million times), looks like this:

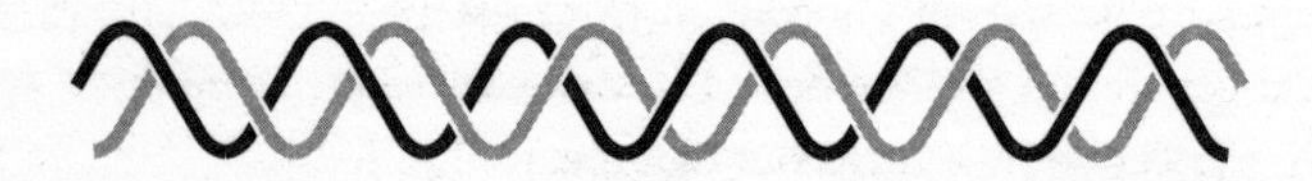

From farther away, those same DNA strands look much smaller, of course, and one can see that each strand is coiled around a protective packaging of histone proteins, which (at a magnification of about 1 million times) looks like this:

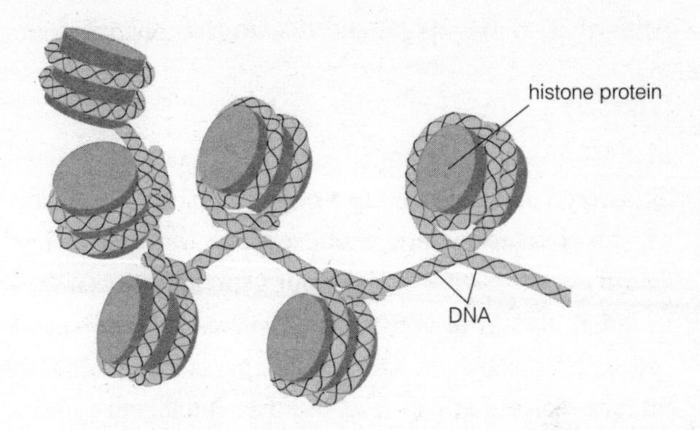

These histones protect the DNA and keep it compact. They also serve as a mediator for gene expression, telling genes when to turn on and off. It's been known for many years that this epigenome ('epi-' is a Latin prefix for 'above' or 'outside') can be altered by the environment and is therefore an important mechanism for gene-environment interaction.

What scientists didn't realise, though, was that changes to the epigenome can be inherited. Prior to 1999, everyone thought that the epigenome was always wiped clean like a blackboard with each new generation.

Not so, discovered Enrico Coen. In the case of the Peloria toadflax flower, a clear alteration to the epigenome had subsequently been passed down through many generations.

And it wasn't just flowers. That same year, Australian geneticists Daniel Morgan and Emma Whitelaw made a very similar discovery in mice. They observed that their batch of genetically identical mice were turning up with a range of different fur colours – differences traced back to epigenetic alterations and passed on to subsequent generations. What's more, they and other researchers discovered that these fur-colour epigenes could be manipulated by something as basic as food. A pregnant yellow mouse eating a diet rich in folic acid or soy milk would be prone to experience an epigenetic mutation producing brown-fur offspring, and even with the pups returning to a normal diet, that brown fur would be passed to future generations.

After that, more epigenetic discoveries piled in one after another:

- In 2004, Washington State University's Michael Skinner discovered that exposure to a pesticide in one generation of rats spurred an epigenetic change that led to low sperm counts lasting at least four generations.
- In 2005, New York University's Dolores Malaspina and colleagues discovered age-related epigenetic changes in human males that can lead to lower intelligence and a higher risk of schizophrenia in children.
- In 2006, London geneticist Marcus Pembrey presented data from Swedish medical records to show that nutritional deficiencies and cigarette smoking in one generation of humans had effects across several generations.
- In 2007, the Institute of Child Health's Megan Hitchins and colleagues reported a link between inherited epigenetic changes and human colon cancer.

Welcome back, Monsieur Lamarck! 'Epigenetics is proving we have some responsibility for the integrity of our genome,' says the Director of Epigenetics and Imprinting at Duke University, Randy Jirtle. 'Before, [we thought that] genes predetermined outcomes. Now [we realise that] everything we do – everything we eat or smoke – can affect our gene expression and that of future generations. Epigenetics introduces the concept of free will into our idea of genetics.'

And that of future generations. This is big, big stuff – perhaps the most important discovery in the science of heredity since the gene.

No one can yet measure the precise implications of these discoveries, because so little is known. But it is already clear that epigenetics is going to radically alter our understanding of disease, human abilities, and evolution. It begins with this simple but utterly breathtaking concept:

Lifestyle can *alter heredity.*

Lamarck was probably not correct about the giraffe in particular, and he was certainly wrong about inherited characteristics being the primary vehicle of evolution. But in its most basic form, his idea that what an individual does in his/her life before having children can change the biological inheritance of those children and their descendants – on this he turns out to have been correct. (And two hundred years ahead of everyone else.) Quietly, biologists have come to accept in recent years that biological heredity and evolution is a lot more intricate than we once thought. The concept of inherited epigenetic changes certainly does not invalidate the theory of natural selection, but it makes it a lot more complicated. It offers not just another mechanism by which species can adapt to changing environments, but also the prospect of an evolutionary process that is more interactive, less random, and runs along several different parallel tracks at the same time. 'DNA is not the be all and end all of heredity,' write geneticists Eva Jablonka and Marion Lamb. 'Information is transferred from one generation to the next by many interacting inheritance systems. Moreover, contrary to current dogma, the variation on which natural selection acts is not always random . . . new heritable variation can arise in response to the conditions of life.'

How do these recent findings impact our understanding of talent and intelligence? We can't yet exactly be sure. But the door of possibility is wide-open. If a geneticist had suggested as recently as the 1990s that a twelve-year-old kid could improve the intellectual nimbleness of his or her future children by studying harder now, that scientist would have been laughed right out of the conference hall. Today, that preposterous scenario looks downright likely:

> Washington, D.C. – New animal research in the February 4 [2009] issue of *The Journal of Neuroscience* shows that a stimulating environment improved the memory of young mice with a memory-impairing genetic defect and also improved the memory of their eventual offspring. The findings suggest that *parental*

> *behaviours that occur long before pregnancy may influence an offspring's well-being*. 'While it has been shown in humans and in animal models that enriched experience can enhance brain function and plasticity, this study is a step forward, suggesting that the enhanced learning behaviour and plasticity can be transmitted to offspring long before the pregnancy of the mother,' said Li-Huei Tsai, PhD, at Massachusetts Institute of Technology and an investigator of the Howard Hughes Medical Institute, an expert unaffiliated with the current study.

In other words, we may well be able to improve the conditions for our grandchildren by putting our young children through intellectual calisthenics now.

What else is possible? Could a family's dedication to athletics in one or more generations induce biological advantages in subsequent generations?

Could a teenager's musical training improve the 'musical ear' of his great-grandchildren?

Could our individual actions be affecting evolution in all sorts of unseen ways?

'People used to think that once your epigenetic code was laid down in early development, that was it for life,' says McGill University epigenetics pioneer Moshe Szyf. 'But life is changing all the time, and the epigenetic code that controls your DNA is turning out to be the mechanism through which we change along with it. Epigenetics tells us that little things in life can have an effect of great magnitude.'

Everything we know about epigenetics so far fits perfectly with the dynamic systems model of human ability. Genes do not dictate what we are to become, but instead are actors in a dynamic process. Genetic expression is modulated by outside forces. 'Inheritance' comes in many different forms: we inherit stable genes, but also alterable epigenes; we inherit languages, ideas, attitudes, but can also change them. We inherit an ecosystem, but can also change it.

Everything shapes us and everything can be shaped by us. The genius in all of us is our built-in ability to improve ourselves and our world.

Epilogue

Ted Williams Field

Parts of the North Park neighborhood of San Diego don't seem to have changed a whole lot since Ted Williams's time. His tiny boyhood home at 4121 Utah Street still stands. Two short blocks away, his old practice baseball field is still there too. They call it 'Ted Williams Field' now. Outside the batting cage are sign-up sheets for Little League. On the sunny afternoon I was there, the field sat empty; no one feverishly hitting baseballs until the threads and skins wore off, no one shagging balls for lunch money. Maybe instead some eleven-year-old kid was inside somewhere practising the cello with all his heart or building a new piece of software that will change the world.

With the field completely empty, it was easier to imagine Ted standing at home plate, yelling at his friend to throw another one – *and harder this time*; to see a few kids standing in the outfield, gloveless, trying to catch the balls but missing most of them. The bat cracks every few seconds, and Ted occasionally mutters, *'That's it, that's it.'* Every time he misses the ball or cracks a foul, he takes note of his stance and his swing. He marks how the ball left the pitcher's hand, how it spun and how it travelled, when he started his swing, and exactly how he moved his shoulders and his hips and his wrists.

I think of my two kids, and I wonder if they'll have the same level of determination with any craft. I wonder if I want them to.

The truth is, I do want my kids to dream big and to never give up. I can't pick their dreams for them and wouldn't dare try. But I can tell them, as my mother and father told me, that any dream is worth having and there's no telling what you can do when you put your mind to it. The only difference between that generation and this one is that my parents were speaking from intuition, faith and experience. I'm speaking from intuition, faith, experience and *science*.

(GxE)

(GxE)

(GxE)

(GxE)

(GxE)

(GxE)

(GxE)

(GxE)

The EVIDENCE

(GxE)

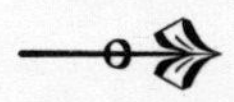

(GxE)

Sources and Notes, Clarifications and Amplifications

BOOK ORIGINS

The notion to pursue a better understanding of talent and giftedness arose from a series of sparks. First, I became intrigued by the 1999 book *Genius Explained* by Michael Howe, which very forcefully took on myths of innate genius and argued that extraordinary abilities can be explained by life's external events. It wasn't entirely persuasive, but it did open my eyes – particularly Howe's deconstruction of the powerful Mozart myth.

Second, during the writing of my previous book on the history of chess, I became intrigued by a number of studies and stories suggesting that even unimaginably great chess minds are constructed over time by emotional drive and extraordinary effort. When the young Alfred Binet studied the great European chess masters of the late nineteenth century (including my great-great-grandfather Samuel Rosenthal), he discovered that they did not – as everyone had assumed – have innately superior visual memories. In fact, their abilities sprouted directly from specific experiential memories that they had created over the years. Later, the Dutch psychologist (and master chess player) Adriaan de Groot, continuing Binet's research, startled the cognitive world with the observation that great chess players also did not actually calculate significantly better or faster than lesser players, nor did they have better memories for raw data than other people. Extraordinary chess players were adept only at the particular skill of seeing chess patterns – the one skill they had spent many thousands of hours studying.

And boy, did they study. Part of the task of understanding high achievement includes a detailed appreciation of the intense and sustained regimen behind it. In that vein, I was struck by something chess columnist Tom Rose wrote about the young Norwegian player Magnus Carlsen. 'He has become a fine player at a very young age. But is that because of exceptional innate talent for chess? Imagine yourself in young Magnus's place. You play in your first tournament aged eight, do well, and get noticed by [a grandmaster] who decides to help teach you. Immediately you believe that you are special, that you have "talent", that you can really shine. This encourages you to work very hard at this game that gets you such agreeable attention . . . [M]ore tournament success and more media attention [encourage] you to work even harder. At first you work at it for 2 or 3 hours a day. By the time you are ten years old it is more like 4 or 5 hours a day.'

This led me to the recently developed science of talent, and to this remark from my near namesake David Shanks, a London psychologist:

> Evidence for the contribution of talent over and above practice has proved extremely elusive . . . [In contrast] evidence is now emerging that exceptional performance in memory, chess, music, sports and other arenas can be fully accounted for on the basis of an age-old adage: practice makes perfect.

'Practice makes perfect' is a terrible phrase, and it invites the obvious question, what about all the people who practise a lot but don't attain high achievement? That's where the work of Anders Ericsson and Neil Charness comes in. Seeing their 1994 paper 'Expert performance – its structure and acquisition' was a revelation. It opened me up to the world of researchers trying to determine precisely how people get good at stuff. There are different degrees of practice, it turns out, and many other elements that go into successful and unsuccessful training, studying, mentoring, etc.

The final spark came after *The Immortal Game* was published. A discussion with the writer Steven Johnson clarified some key issues; a second conversation with the writer Cathryn Jakobson Ramin prompted Ramin to send me a provocative editorial, 'The Sky's the

Limit,' from the September 16, 2006, issue of *New Scientist.* That piece very succinctly suggested that maybe it was time for a wholesale reevaluation of the notion of talent and alerted me to the critical work of Carol Dweck on the questions of mind-set and motivation.

From there, I dug up and read through a truckload of journal articles and books, eventually realising that I was bouncing between two very separate scientific worlds: the study of genetics and the study of talent/achievement. Each had undergone great transformations in recent years that scientists themselves were still struggling to articulate – with frankly very limited success. My ambitious goal became to try to somehow bridge these two worlds and to distil it all into a new lingua franca, adopting helpful new phrases and metaphors that scientists could share with teachers, journalists, politicians, and so on. And so the odyssey began . . .

INITIAL SOURCES

Binet, Alfred. *Mnemonic Virtuosity: A Study of Chess Players*, 1893. Translated by Marianne L. Simmel and Susan B. Barron. Journal Press, 1966.

de Groot, Adrianus Dingeman. *Thought and Choice in Chess.* Walter de Gruyter, 1978.

Elliot, Andrew J., and Carol S. Dweck, eds. *Handbook of Competence and Motivation.* Guilford Publications, 2005.

Ericsson, K. Anders, and Neil Charness. 'Expert performance – its structure and acquisition'. *American Psychologist* 49, no. 8 (August 1994): 725–47.

Ericsson, K. Anders, Neil Charness, Paul J. Feltovich, and Robert R. Hoffman, eds. *The Cambridge Handbook of Expertise and Expert Performance.* Cambridge University Press, 2006.

Howe, Michael. *Genius Explained.* Cambridge University Press, 1999.

New Scientist Editorial Board. 'The Sky's the Limit'. *New Scientist*, September 16, 2006.

Ridley, Matt. *Nature via Nurture.* HarperCollins, 2003.

Rose, Tom. 'Can "old" players improve all that much?' Published on the Chessville.com Web site.

Shanks, D. R. 'Outstanding performers: created, not born? New results on nature vs. nurture'. *Science Spectra* 18 (1999).

INTRODUCTION: THE KID

CHAPTER NOTES

5 **'I remember watching one of his home runs from the bleachers of Shibe Park,' John Updike wrote:** Updike, 'Hub Fans Bid Kid Adieu', p. 112.

5 **'Ted just had that natural ability,' said Hall of Fame second baseman Bobby Doerr:** Nowlin and Prime, *Ted Williams*, p. 34.

5 **'Ted Williams sees more of the ball than any man alive,' Ty Cobb once remarked:** *USA Today* Editors, 'In every sense, Williams saw more than most'.

In the same vein, former Cincinnati pitcher Johnny Vander Meer adds:

> The first time I saw Ted Williams was at an exhibition game at Plant Field in Tampa. He was a rookie with the Boston club and I was with Cincinnati. He was the last man up in the game, in the ninth, and I was working the last two or three innings. Ted took a third strike.
>
> The game was over and I was walking off the field, he came up to me and asked, 'Did you make the ball spin the other way? Did you turn it over?' The ball was low and inside; it was a sinker. 'I sure did,' I said. I had turned my hand over – my wrist over – and put a reverse spin on the ball. Bucky Walters was standing right close by and I said to Bucky, 'That guy sees which way the stitches are turning! He ought to prove a pretty good hitter.' Hell yeah, he saw the stitches! Or he wouldn't have asked me if I'd turned the ball over. (From Nowlin and Prime, *Ted Williams*, p. 34.)

5 **'a lot of bull':** Montville, *Ted Williams*, p. 26.

6 **'His whole life was hitting the ball,' recalled a boyhood friend.**

The friend is Roy Engle. Two separate quotes are stitched together here. (Nowlin and Prime, *Ted Williams*, pp. 6–8.)

—→ In 1991, biographer Bill Nowlin was in San Diego to attend the renaming of Williams's old practice field as 'Ted Williams Field'. Nowlin spent time with many people who had known Ted from childhood. 'I had wondered if there was any indication when Ted was a boy that he would turn into the great player that he was; any indication that he was to be the anointed one. While obviously a good player, apparently nothing stood out that marked Ted above the other good players around the neighbourhoods at that time. As one old friend put it, "He was good all right, but it wasn't until about 15 that he started shooting ahead of the rest of us. After that, there was just no stopping him."' (Nowlin, *The Kid*, p. 210.)

6 **At San Diego's old North Park field:** Edes, 'Gone'.

7 **Frank Shellenback noticed that his new recruit:** Nowlin and Prime, *Ted Williams*, p. 14.

8 **'He discussed the science of hitting':** Nowlin and Prime, *Ted Williams*, p. x.

8 **'pitchers figure out [batters'] weaknesses,' said Cedric Durst:** Nowlin and Prime, *Ted Williams*, p. 13.

9 **If humans were fruit flies, with a new generation appearing every eleven days, we might be tempted to chalk it up to genetics and rapid evolution.**
—◦⋙ A single fly's random genetic mutation can spread into a whole community in a matter of months. Scientists have demonstrated this many times over, breeding gladiator flies, super-memory flies, flightless flies, and so on.

10 **'unactualised potential':** This term comes from Ceci, Rosenblum, de Bruyn, and Lee, 'A Bio-Ecological Model of Intellectual Development', p. 304.

10 **'We have no way of knowing how much unactualised genetic potential exists':** Ceci, Rosenblum, de Bruyn, and Lee, 'A Bio-Ecological Model of Intellectual Development', p. xv.

10 **This new paradigm does not herald a simple shift from 'nature' to 'nurture'. Instead, it reveals how bankrupt the phrase 'nature versus nurture' really is and demands a whole new consideration of how each of us becomes us.**
—◦⋙ Among the many scientists calling for this new consideration are Penn State geneticist Gerald E. McClearn. 'For most of the past century,' he writes, 'the evidence has been clear that a more collaborative model of coaction and interaction of genetic and environmental agencies is more appropriate.' (Gerald E. McClearn, 'Nature and nurture', pp. 124–30.)

As this book was going to press, Mark Blumberg made me aware of a new paper arguing that the phrase 'nature versus nurture' should be forever banished. The citation: Spencer, J. P., M. S. Blumberg, R. McMurray, S. R. Robinson, L. K. Samuelson, and J. B. Tomblin. 'Short arms and talking eggs: why we should no longer abide the nativist-empiricist debate'. (*Child Development Perspectives*, July 2009.)

CHAPTER 1:
GENES 2.0 – HOW GENES REALLY WORK

PRIMARY SOURCES

My understanding of how genes work and traits develop comes from hundreds of books and articles. The most important (in alphabetical order) are as follows:

Bateson, Patrick, and Paul Martin. *Design for a Life: How Biology and Psychology Shape Human Behavior*. Simon & Schuster, 2001.

Bateson, Patrick, and Matteo Mameli. 'The innate and the acquired: useful clusters or a residual distinction from folk biology?' *Developmental Psychobiology* 49 (2007): 818–31.

Godfrey-Smith, Peter. 'Genes and Codes: Lessons from the Philosophy of Mind?' In *Biology Meets Psychology: Constraints, Conjectures, Connections*, edited by V. Q. Hardcastle. MIT Press, 1999, 305–31.

Gottlieb, Gilbert. 'On making behavioral genetics truly developmental'. *Human Development* 46 (2003): 337–55.

Griffiths, Paul. 'The Fearless Vampire Conservator: Phillip Kitcher and Genetic Determinism'. In *Genes in Development: Rereading the Molecular Paradigm*, edited by E. M. Neumann-Held and C. Rehmann-Sutter. Duke University Press, 2006.

Jablonka, Eva, and Marion J. Lamb. *Evolution in Four Dimensions*. MIT Press, 2005.

Johnston, Timothy D., and Laura Edwards. 'Genes, interactions, and the development of behavior'. *Psychological Review* 109, no. 1 (2002): 26–34.

McClearn, Gerald E. 'Nature and nurture: interaction and coaction'. *American Journal of Medical Genetics* 124B, no. 1 (2004): 124–30.

Meaney, Michael J. 'Nature, nurture, and the disunity of knowledge'. *Annals of the New York Academy of Sciences* 935 (2001): 50–61.

Moore, David S. *The Dependent Gene: The Fallacy of 'Nature vs. Nurture'*. Henry Holt, 2003.

Oyama, Susan, Paul E. Griffiths, and Russell D. Gray. *Cycles of Contingency: Developmental Systems and Evolution*. MIT Press, 2003.

Pigliucci, Massimo. *Phenotypic Plasticity: Beyond Nature and Nurture*. Johns Hopkins University Press, 2001.

Ridley, Matt. *Nature via Nurture*. HarperCollins, 2003.

Rutter, Michael, Terrie E. Moffitt, and Avshalom Caspi. 'Gene-environment interplay and psychopathology: multiple varieties but real effects'. *Journal of Child Psychology and Psychiatry* 47, no. 3/4 (2006): 226–61.

Turkheimer, Eric. 'Three laws of behavior genetics and what they mean'. *Current Directions in Psychological Science* 9, no. 5 (October 2000): 160–64.

While it would be impossible to further rank the above works in terms of their brilliance or general importance, I must give special credit to Matt Ridley's *Nature via Nurture* for its importance in laying down a basic new foundation of knowledge of gene-environment interaction. Which does not, of course, mean that Ridley should get blamed for any of my silly mistakes . . .

CHAPTER NOTES

14 **And to think [I'm] the cause of it:** Chase and Winter, 'The Sopranos: Walk Like a Man', May 6, 2007.

14 **The irony is that as America equalises the [environmental] circumstances:** Herrnstein and Murray, *The Bell Curve*, p. 91.

→ There's also this gem: 'Universal college education cannot be. Most people are not smart enough to profit from an authentic college education.' (Murray and Seligman, 'As the Bell Curves'.)

14 **'There are no genetic factors that can be studied independently of the environment.'**

→ He uses 'phenotype' instead of 'a trait'. I substituted so as not to distract the reader. Here's the original quote: 'There are no genetic factors that can be studied independently of the environment, and there are no environmental factors that function independently of the genome. Phenotype emerges only from the interaction of gene and environment.' Meaney continues: 'The search for main [direct] effects is a fool's errand. In the context of modern molecular biology, it is a quest that is without credibility.' (Meaney, 'Nature, Nurture, and the Disunity of Knowledge', pp. 50–61.)

14–15 **We've all been taught that we inherit complex traits like intelligence straight from our parents' DNA in the same way we inherit simple traits like eye colour. This belief is continually reinforced by the popular media.**

A few examples:

'An organism's physiology and behaviour are dictated largely by its genes,' the *Economist* declared in 1999. (Griffiths, 'The Fearless Vampire', p. 4.)

In 2005, *Scientific American* affirmed: 'Even such abstract qualities as personality and intelligence are coded for in our genetic blueprint.' (Gazzaniga, 'Smarter on Drugs', p. 32.)

On November 11, 2008, as the writing of this book was drawing to a close, the *New York Times* published a remarkable piece by Carl Zimmer acknowledging a revolutionary new understanding of genes. Some key excerpts:

> The familiar double helix of DNA no longer has a monopoly on heredity. Other molecules clinging to DNA can produce striking differences between two organisms with the same genes. And those molecules can be inherited along with DNA . . . It turns out, for example, that several different proteins may be produced from a single stretch of DNA . . . It turns out that the genome is also organised in another way, one that brings into question how important genes are in heredity. Our DNA is studded with millions of proteins and other molecules, which determine which genes can produce transcripts and which cannot. New cells inherit those molecules along with DNA. In other words, heredity can flow through a second channel. (Zimmer, 'Now: The Rest of the Genome'.)

Still, the online *New York Times* health guide, under the heading 'Genetics', crudely states: 'It is common knowledge that a person's appearance – height, hair colour, skin colour, and eye colour – are determined by genes. Mental abilities and natural talents are also affected by heredity, as is the susceptibility to acquire certain diseases.'

15 **Think of your own genetic makeup:** Friend, 'Blueprint for Life', p. D 01.

15 **Gregor Mendel demonstrated that basic traits:** Field Museum, 'Gregor Mendel: Planting the Seeds of Genetics'.

15 **Mendel had proved the existence of genes – and seemed to prove that genes alone determined the essence of who we are. Such was the unequivocal interpretation of early-twentieth-century geneticists.**
→ Pitzer College's David S. Moore provides a nice capsule history of genetic determinism from the time of Mendel:

> The idea that genetic factors might be able to determine the form of biological and psychological traits has been with us since the beginning of modern theorising about genes. Although Gregor Mendel did not use

> the word genes to name the 'heritable factors' that he inferred must be responsible for observed variations in his experimental pea plants, the notion of a deterministic 'germ plasm' had appeared in several late 19th century writings on biology – most notably in the work of August Weismann – and because of the close conceptual similarity between Mendel's 'heritable factors' and Weismann's deterministic 'germ plasm', it is little wonder that just a few decades later, Mendel's factors came to be thought of as deterministic 'genes'. T. H. Morgan's early 20th century discovery that genes are located on chromosomes eventually led to the development of the modern gene theory, which holds that genes are responsible for the development of inherited traits; this conclusion was based on the finding that the presence of particular genetic factors is highly correlated with the presence of particular traits. But even though such correlations do not support the contention that genes operate deterministically, modern gene theory nonetheless retained the genetic determinism that 19th century 'germ plasm' theorists relied on to explain the intergenerational transmission of evolutionarily adaptive characteristics. This sort of conceptualisation continued to inform theoretical biology well past the middle of the 20th century, as biologists embraced Francois Jacob and Jacques Monod's operon model of how genes regulate development. (Moore, 'Espousing interactions and fielding reactions', p. 332.)

Moore also notes that Johannsen recognised that development *was* a factor, and that they were ignoring development with their genes-only approach. (Moore, *The Dependent Gene*, p. 167.)

15 'It's in the genes,' we say.

—⊸ What makes Michael Phelps such an outstanding swimmer? It's 'all about gene pool,' quips syndicated sports columnist Rob Longley. 'Phelps [has been] blessed with so many gifts, he is nothing short of a freak of nature.' (Long-ley column.)

15 over the last two decades Mendel's ideas have been thoroughly upgraded – so much so that one large group of scientists now suggests that we need to wipe the slate clean and construct an entirely new understanding of genes.

—⊸ Ironically, as this sweeping new view of how genes work has emerged, it has received little public attention. Front-page headlines still trumpet advances in gene splicing, genome mapping, gene testing, cloning, and so on. The result has been a growing public disconnect between genetic understanding and genetic reality. The public has got the impression that the answer to almost every question about our health and well-being can be found in our genome. The reality is a lot more nuanced.

16 Not all of the interactionists' views have yet been fully accepted.

—⊸ This book is not a dispassionate presentation of all scientific points of view. Instead, it embraces the arguments of the Interactionists, whose views I came to trust most after much reading, conversation, and consideration.

One brief description of the running disagreement can be found in Johnson and Karmiloff-Smith, 'Neuroscience Perspectives on Infant Development', which may be accessed online via Google Books (go to 'Contents', and click on page 121).

Another comes from Patrick Bateson and Matteo Mameli:

> Many authors writing today suppose that innateness has something to do with genes (e.g., Tooby & Cosmides, 1992; Plotkin, 1997; Chomsky, 2000; Fodor, 2001; Pinker, 1998, 2002; Miller, 2000; Baron-Cohen, 2003; Buss, 2003; Marcus, 2003; Marler, 2004). In some cases, this supposition is based on imprecise ways of thinking about the role of genes in development. To argue, for instance, that a phenotype is innate if and only if genes and nothing but genes are required for its development is too simplistic. No phenotype is such that only genes are needed for its development, since an interplay between the organism and its environment is required at all stages of development. (Bateson and Mameli, 'The innate and the acquired', p. 819.)

16 'The popular conception of the gene as a simple causal agent is not valid,' declare geneticists Eva Jablonka and Marion Lamb.

They add: '[Geneticists now] recognise that whether or not a trait develops does not depend, in the majority of cases, on a difference in a single gene. It involves interactions among many genes, many proteins and other types of molecule[s], and the environment in which an individual develops.'

Also: 'The idea that there is a gene for adventurousness, heart disease, obesity, religiosity, homosexuality, shyness, stupidity, or any other aspect of mind or body has no place on the platform of genetic discourse.' (Jablonka and Lamb, *Evolution in Four Dimensions*, pp. 6–7.)

16 This obliterates the long-standing metaphor of genes as blueprints with elaborate predesigned instructions for eye colour, thumb size, mathematical quickness, musical sensitivity, etc.

→ Deploying the right metaphor is everything in the communication and understanding of science. In the case of genetics, our metaphors have sadly led us astray. 'There is no clear, technical notion of "information" in molecular biology,' writes biologist and philosopher Sahotra Sarkar. 'It is little more than a metaphor that masquerades as a theoretical concept and . . . leads to a misleading picture of possible explanations in molecular biology.'

Today's popular understanding of genes, heredity, and evolution is not just crude; it is profoundly misleading. It may *feel* true, thanks to the elegance of the 'blueprint' and 'code' metaphors, and thanks to the lack of a cogent dissent. But from the vantage of twenty-first-century scientific understanding, any brand of genetic determinism obscures more than it enlightens. We've created a thick, semipermanent veil that shrouds the more interesting, and more hopeful, reality.

'What we need here,' writes John Jay College's Susan Oyama (a leader in the dynamic systems movement), 'is the stake-in-the-heart move, and the heart is the notion that some influences are more equal than others, that form, or its modern agent, information, exists before the interactions in which it appears and must be transmitted to the organism either through the genes or by the environment.' (Oyama, *The Ontogeny of Information*, p. 27.)

16 genes – all twenty-two thousand of them – are more like volume knobs and switches.

→ This is my attempt to come up with a metaphor that will resonate and that accurately captures the dynamic quality of genes.

16 Estimates of the actual number of genes vary.

Although the completion of the Human Genome Project was celebrated in April 2003 and sequencing of the human chromosomes is essentially 'finished', the exact number of genes encoded by the genome is still unknown. October 2004 findings from the International Human Genome Sequencing Consortium, led in the United States by the National Human Genome Research Institute (NHGRI) and the Department of Energy (DOE), reduce the estimated number of human protein-coding genes from 35,000 to only 20,000–25,000, a surprisingly low number for our species. Consortium researchers have confirmed the existence of 19,599 protein-coding genes in the human genome and identified another 2,188 DNA segments that are predicted to be protein-coding genes. In 2003, estimates from gene-prediction programmes suggested there might be 24,500 or fewer protein-coding genes. The Ensembl genome-annotation system estimates them at 23,299. (Human Genome Project, 'How Many Genes Are in the Human Genome?')

Also: New data 'threaten to throw the very concept of "the gene" – either as a unit of structure or as a unit of function – into blatant disarray.' (Keller, *The Century of the Gene*, p. 67.)

16 Many of those knobs and switches can be turned up/down/on/off at any time – by another gene or by any minuscule environmental input. This flipping and turning takes place constantly.

> Experiential factors are now known to influence gene expression through several mechanisms, including (but not limited to) those involving the actions of steroid hormones . . . For example, testosterone levels change as a function of sexual experience, and hormones like testosterone are known to be able to diffuse across both cellular and nuclear membranes where – once they have been bound by specific receptors – they can bind with DNA to regulate gene expression. (Moore, 'Espousing interactions and fielding reactions', p. 340.)

17 this process of gene-environment interaction drives a unique developmental path for every unique individual.

'The process of GxE acting over a lifetime may be the key to understanding much of human complex trait variability.' (Brutsaert and Parra, 'What makes a champion?', p. 110.)

18 This may sound crazy at first, because of how thoroughly we've been indoctrinated with Mendelian genetics. The reality turns out to be much more complicated – even for pea plants.

—⊸ Mendel's pea-plant example has a built-in logical flaw: by assuring a consistent environment, it eliminates any visible environmental impact on heredity. When the environment is perfectly consistent from plant to plant, it does indeed appear that genes single-handedly determine heredity. This is akin to throwing dice, but instead of rolling two dice at once, keeping one of them permanently on 6. The second die is always going to determine the total.

18 Many scientists have understood this much more complicated truth for years but have had trouble explaining it to the general public. It is indeed a lot harder to explain than simple genetic determinism.

—∘⇛ In a 2009 essay for the *New York Times Magazine*, Steven Pinker writes: 'For most . . . traits, any influence of the genes will be *probabilistic*. Having a version of a gene may change the odds, making you more or less likely to have a trait, all things being equal, but as we shall see, the actual outcome depends on a tangle of other circumstances as well.' (Italics mine.)

While this is an important acknowledgement that most genes do not determine traits directly, the use of the word 'probabilistic' is crude and troublesome in two ways: First, it gives a *new* wrong impression about how genes work – making them sound like dice. Second, it misses a critical opportunity to help the general public understand genetic expression and gene-environment interaction.

The term 'probabilistic' is meant to convey the understanding that most specific gene variants (alleles) do not guarantee certain outcomes. That much is true.

But the term goes much further. It also conveys the strong sense that a certain gene creates a specific probability that a person will develop a certain trait. That is very misleading – as Pinker himself demonstrates.

To explore the current state of genetics, Pinker had his own DNA analysed. Among other things, it was revealed that he had the T version of a gene called *rs2180439 SNP*. As it turns out, 80 per cent of men with the T version of this gene are bald. Pinker has a head full of curly grey hair. 'Something strange happens when you take a number representing the proportion of people in a sample and apply it to a single individual,' he writes. 'The first use of the number is perfectly respectable as an input into a policy that will optimise the costs and benefits of treating a large similar group in a particular way. But the second use of the number is just plain weird.'

Exactly. And that is also, in my opinion, why it is a bad idea to use the word 'probabilistic' to describe the nature of genes. Genes don't always lead to certain outcomes, because they are involved in a complex gene-environment dynamic. For the exact same reason, genes also don't create a specific probability of an outcome.

My argument with the term 'probabilistic' is not an argument against population genetics research. Such studies can be darn useful in setting medical policy, as Pinker suggests. But such studies should not drive our descriptive terminology for genes and how they work. (Quotes from Pinker, 'My Genome, My Self'.)

18 Proteins are large, specialised molecules that help create cells, transport vital elements, and produce necessary chemical reactions.

From the online Genetics Home Reference guide:

What are proteins and what do they do?

Proteins are large, complex molecules that play many critical roles in the body. They do most of the work in cells and are required for the structure, function, and regulation of the body's tissues and organs. Proteins are made up of hundreds or thousands of smaller units called amino acids, which are attached to one another in long chains. There are 20 different types of amino acids that can be combined to make a protein. The sequence of amino acids determines each protein's unique 3-dimensional structure and its specific function. Proteins can be described according to their large range of functions in the body, listed in alphabetical order:

Examples of protein functions

Antibody: Antibodies bind to specific foreign particles, such as viruses and bacteria, to help protect the body.

Enzyme: Enzymes carry out almost all of the thousands of chemical reactions that take place in cells. They also assist with the formation of new molecules by reading the genetic information stored in DNA.

Messenger: Messenger proteins, such as some types of hormones, transmit signals to coordinate biological processes between different cells, tissues, and organs.

Structural component: These proteins provide structure and support for cells. On a larger scale, they also allow the body to move.

Transport/storage: These proteins bind and carry atoms and small molecules within cells and throughout the body.

19 This explains how every brain cell and hair cell and heart cell in your body can contain *all* of your DNA but still perform very specialised functions.

Lawrence Harper writes:

Every cell inherits a full nuclear complement of DNA. That is, all cells in the organism have the same potential. In the presence of appropriate external conditions, what underlies the development of multicellular organisms is a progressive, differential production (*expression*) of certain subsets of this genetic potential in different tissues . . . The features of each tissue type are thus determined by the pattern of gene expression, the genes in the cells that are 'turned on' or 'off' or show distinctive rates of production of gene products. (Harper, 'Epigenetic inheritance and the intergenerational transfer of experience', p. 344.)

20 **'Development is chemistry':** Brockman, 'Design for a Life: A Talk with Patrick Bateson'.

20–21 All of this means that, on their own, most genes cannot be counted on to directly produce specific traits. They are active participants in the developmental process and are built for flexibility. Anyone seeking to describe them as passive instruction manuals is actually minimising the beauty and power of the genetic design.

Lawrence Harper writes:

> Of particular relevance to the understanding of behavioural ontogeny is the fact that, in the process of development, cellular gene expression can be stably altered in response to conditions outside the organism to permit it to adapt to its environment. That is, not only do cells differentiate (specialise in function) in response to external signals, but once so differentiated, their subsequent functional activity as, for example, nerves or glandular tissue, also can be modified at the molecular level. Probably the most obvious example of such altered activity of specialised cells is the development of immunity to pathogens. (Harper, 'Epigenetic inheritance and the intergenerational transfer of experience', p. 345.)

21 **'Even in the case of eye colour,' says Patrick Bateson, 'the notion that the relevant gene is *the* [only] cause is misconceived, because [of] all the other genetic and environmental ingredients.' (Italics mine).** Bateson, 'Behavioral Development and Darwinian Evolution', p. 149.

—⊸ A taste of the complexities behind eye colour, from three different sources:

> Iris colour was one of the first human traits used in investigating Mendelian inheritance in humans. Davenport and Davenport (1907) outlined what was long taught in schools as a beginner's guide to genetics, that brown eye colour is always dominant to blue, with 2 blue-eyed parents always producing a blue-eyed child, never one with brown eyes. As with many physical traits, the simplistic model does not convey the fact that eye colour is inherited as a polygenic, not as a monogenic, trait (Sturm and Frudakis, 2004). Although not common, 2 blue-eyed

parents can produce children with brown eyes. (McKusick, 'Eye Color 1'.)

* * *

Human iris colour is a quantitative, multifactorial phenotype that exhibits quasi-Mendelian inheritance . . . To identify genetic features for best-predicting iris colour, we selected sets of SNPs by parsing P values among possible combinations . . . These results confirm that OCA2 is the major human iris colour gene and suggest that using an empirical database-driven system, genotypes from a modest number of SNPs within this gene can be used to accurately predict iris melanin content from DNA. (Frudakis, Terravainen, and Thomas, 'Multilocus OCA2 genotypes specify human iris colors', pp. 3311–26.)

* * *

The highest association for blue/nonblue eye colour was found with three OCA2 SNPs . . . The TGT/TGT diplotype found in 62.2% of samples was the major genotype seen to modify eye colour, with a frequency of 0.905 in blue or green compared with only 0.095 in brown eye colour. This genotype was also at highest frequency in subjects with light brown hair and was more frequent in fair and medium skin types, consistent with the TGT haplotype acting as a recessive modifier of lighter pigmentary phenotypes. (Duffy et al., 'A three-single-nucleotide polymorphism haplotype in intron 1 of OCA2 explains most human eye-color variation', p. 241.)

21 **Single-gene diseases do exist and account for roughly 5 per cent of the total disease burden in developed countries:** Khoury, Yang, Gwinn, Little, and Flanders, 'An epidemiological assessment of genomic profiling for measuring susceptibility to common diseases and targeting interventions', pp. 38–47; Hall, Morley, and Lucke, 'The prediction of disease risk in genomic medicine'.

Susan Brooks Thistlethwaite adds:

Genetics is not merely a matter of single gene disorders or single gene traits, such as flower colour and pod shape in Mendel's pea plants. Mendelian genetics is about single gene disorders [that] occur in only 3 per cent of all individuals born alive . . .

Human inheritance is much more complicated. Most conditions are polygenic (involve many genes), and their expression depends on gene-gene and environment-gene interactions. (Thistlethwaite, *Adam, Eve, and the Genome*, p. 70.)

21 **'A disconnected wire can cause a car to break down':** Oyama, Griffiths, and Gray, *Cycles of Contingency*, p. 157.

22 **'Genes store information coding for the amino acid sequences of proteins,' explains Bateson. 'That is all':** Bateson, *Design for a Life*, p. 66.

Similar statement: 'All the genes can code for, if they code for anything, is the primary structure (amino acid sequence) of a protein.' (Godfrey-Smith, 'Genes and Codes', p. 328.)

22 **One of the most striking early hints of the new understanding of development as a dynamic process emerged in 1957.**

→ There were much earlier hints. 'For most of the past century,' says Penn State geneticist Gerald E. McClearn, 'the evidence has been clear that a more collaborative model of coaction and interaction of genetic and environmental agencies is more appropriate. Even in the pell-mell pursuit of Mendelian phenomena in the post-rediscovery enthusiasm at the beginning of the last century, examples of the interdependence of genetic and environmental influences surfaced. One well-known early example is that of Krafka [1920], who showed that the effect of the bar-eyed genotype (now known to be a duplication) on eye facet number of *Drosophila* is strikingly dependent on the temperature at which the flies are maintained.' (McClearn, 'Nature and nurture', p. 124.)

23 **heights of Japanese children:** Greulich, 'A comparison of the physical growth and development of American-born and native Japanese children', p. 304.

23 **Greulich didn't realise this at the time, but it was a perfect illustration of how genes really work: not dictating any predetermined forms or figures, but interacting vigorously with the outside world to produce an improvised, unique result.**

→ Two excellent summaries from two of the top figures in the field of gene-environment interaction:

> A key feature of gene expression is that it can be altered in a reversible way by extra-cellular signals and by environmental influences. Although DNA starts off the causal chain, what really matters is the expression of the genes (in terms of messenger RNA). There are no genetic effects

without this expression. (Rutter, Moffitt, and Caspi, 'Gene-environment interplay and psychopathology', p. 229.)

* * *

Individual genes and their environments interact to initiate a complex developmental process that determines adult personality. Most characteristic of this process is its interactivity: Subsequent environments to which the organism is exposed depend on earlier states, and each new environment changes the developmental trajectory, which affects future expression of genes, and so forth. Everything is interactive, in the sense that no arrows proceed uninterrupted from cause to effect; any individual gene or environmental event produces an effect only by interacting with other genes and environments. (Turkheimer, 'Three laws of behavior genetics and what they mean', p. 161.)

23 **in truth human height has fluctuated dramatically over time.** This from height anthropologist Richard Steckel: 'We have 1200 years of adult male height trends in Northern Europe that show that height was greatest in the early middle ages, when there was a warmer climate, and reached a minimum in the Little Ice Age of the 17th and 18th centuries.' (Steckel, 'Height, Health, and Living Standards Conference Summary', p. 13.)

Also: American and British teenagers were six inches taller, on average, than their predecessors a century earlier. (Ceci, Rosenblum, DeBruyn, and Lee, 'A Bio-Ecological Model of Intellectual Development'.)

23 ***The New Yorker*'s Burkhard Bilger:** Bilger, 'The Height Gap'. A few more excerpts from Bilger's piece:

Though climate still shapes musk oxen and giraffes – and a willowy Inuit is hard to find – its effect on industrialised people has almost disappeared. Swedes ought to be short and stocky, yet they've had good clothing and shelter for so long that they're some of the tallest people in the world. Mexicans ought to be tall and slender. Yet they're so often stunted by poor diet and diseases that we assume they were born to be small.

Biologists say that we achieve our stature in three spurts: the first in infancy, the second between the ages of six and eight, the last in adolescence. Any decent diet can send us sprouting at these ages, but take away any one of forty-five or fifty essential nutrients and the body stops growing. ('Iodine deficiency alone can knock off ten centimetres and fifteen I.Q. points,' one nutritionist told me.)

Steckel, after his work on slaves, went on to Union soldiers and Native Americans. (The men of the northern Cheyenne, he found, were the tallest people in the world in the late nineteenth century: well nourished on bison and berries, and wandering clear of disease on the high plains, they averaged nearly five feet ten.) Then he enlisted anthropologists to gather bone measurements dating back ten thousand years. In both Europe and the Americas, he discovered, humans grew shorter as their cities grew larger. The more people clustered together, the more pest-ridden and poorly fed they became. Heights also fell in synch with global temperatures, which reached a nadir during the Little Ice Age of the seventeenth century.

Around the time of the Civil War, Americans' heights predictably decreased: Union soldiers dropped from sixty-eight to sixty-seven inches in the mid-eighteen-hundreds, and similar patterns held for West Point cadets, Amherst students, and free blacks in Maryland and Virginia. By the end of the nineteenth century, however, the country seemed set to regain its eminence. The economy was expanding at a dramatic rate, and public-hygiene campaigns were sweeping the cities clean at last: for the first time in American history, urbanites began to outgrow farmers.

In personal correspondence, Patrick Bateson warns: '[Don't] overstate your case. Differences in genes can be correlated with a difference in behaviour or morphology. Not everyone will reach the same height if they are all given a superb diet. Pygmies, for example, produce less growth hormone or, in the case of other populations (the phenotype seems to have evolved at least five times in different parts of the world), are less receptive to growth hormone.'

24 'Maze-dull' rats, which had consistently tested poorly in those same mazes, making an average of 40 per cent more mistakes.

—⊸ This second group consistently stumbled through the same maze over and over again without remembering or learning, making an average of 40 per cent more mistakes than the smarter group. They seemed obviously dumber than the Maze-bright strain, possessing an apparently inferior set of intelligence genes.

26 'a classic example of gene-environment interaction': McClearn, 'Genetics, Behavior and Aging', p. 11.

26 **temperature surrounding turtle and crocodile eggs determined their gender:** Bateson, 'Behavioral Development and Darwinian Evolution', p. 52.

26 **In 1972, Harvard biologist Richard Lewontin supplied a critical clarification that helped his colleagues understand GxE.**

Paolo Vineis, chair of Environmental Epidemiology, Imperial College, London, explains:

> This issue was clarified in an important paper by Richard Lewontin many years ago, but it is still a matter of confusion. The main idea of Lewontin's paper is that when we evaluate gene-environment interactions we use the 'analysis of variance' paradigm, that is, we try to combine the two main effects (genes versus environment), plus their interactive term, in a linear model. Causal models presuppose a linear combination of factors as the base line, variances are then computed and the role of the two main effects (or their interaction) is apportioned accordingly. But, Lewontin argues, the analysis-of-variance approach is misleading. There is no theoretical justification for the presumption of a linear explanation (this is done for the sake of simplicity but does not correspond to any reasonable biological reason). By contrast, all the experiments done with, for example, *Arabidopsis* (a plant) or *Drosophila* (based for example on radiation-induced mutations) show that mutations cause a change in what is called the 'norm of reaction', that is, the ability of the organism to react to different environmental conditions. The way in which the mutant strain will react, say, to different temperatures, is not predictable if the environmental conditions are not specified. Usually what happens is 'canalisation', that is, under 'normal' conditions there is a certain norm of reaction that is the same for the wild type and the mutants, whereas in changing environments the wild type and the mutant differ in the norm of reaction. *What this suggests is that in at least some cases a nonlinear explanation is going to be required. In practical terms, it means that all attempts to explain disease on the basis of either the environment or genes (or their interaction) are in fact doomed to fail, because two organisms with different gene variants will have exactly the same response in a normal environment, and a totally different response in an abnormal environment.* (Italics mine.) (Vineis, 'Misuse of genetic data in environmental epidemiology', pp. 164–65. The paper Vineis is referring to is Lewontin, 'The analysis of variance and the analysis of causes'.)

27 **'the way genes and environments interact dialectically to generate an organism's appearance and behaviour':** Pigliucci, 'Beyond nature and nurture', pp. 20–22.

27 **'the individual animal starts its life with the capacity to develop in a number of distinctly different ways':** Bateson and Martin, *Design for a Life*, pp. 102–3.

'Everything we have learned about molecular biology has shown that gene activity is regulated by the intracellular environment,' explains McGill's Michael Meaney. He continues:

> The intracellular environment is a function of the genetic make-up of the cell and the extracellular environment (e.g. hormones released by endocrine organs, cytokines from the immune system, neurotransmitters from neurons, nutrients derived from food) [which is] also influenced by the environment of the individual. Neurotransmitter and hormonal activity is profoundly influenced, for example, by social interactions, which lead to effects on gene activity. (Meaney, 'Nature, nurture, and the disunity of knowledge', p. 52.)

28 **Your life is interacting with your genes.**

—◆➾ If genes are merely the bricklayers, where's the foreman? Where's the architect?

Amazingly, there is no architect. Like ant colonies, galaxies, and other complex emergent systems, the human body is a dynamic assembly abiding by certain strict laws of science but not following any master set of instructions. The outcome is a function of the ingredients and the process.

The University of Virginia's Eric Turkheimer explains it this way: 'Individual genes and their environments interact to initiate a complex developmental process that determines adult personality. Most characteristic of this process is its interactivity. Subsequent environments to which the organism is exposed depend on earlier states, and each new environment changes the developmental trajectory, which affects future expression of genes, and so forth. Everything is interactive, in the sense that no arrows proceed uninterrupted from cause to effect; any individual gene or environmental event produces an effect only by interacting with other genes and environments.'

The point here is not to suggest that every person has exactly the same biological advantages or limits, or exactly the same potential. We clearly do not. But understanding each person's true potential is not something we'll ever be able to do from a genetic snapshot. Too many developmental factors matter too much. When it comes to complex traits like intelligence and talent, we need to drop casual use of the word 'innate' and instead strive to understand as much as we can about the gene-influenced, environment-mediated process called human development.

While the scientific use of the word 'innate' is still under intense discussion among biologists, it's clear enough that its popular use to refer to fixed, built-in, predetermined causes of complex traits is simply no longer supportable. It has become obsolete.

Like the popular use of the word 'genes,' it is a mere stand-in for things we don't understand about how we become who we are, a shorthand for the rich and enigmatic incubator of temperament, inclinations, and abilities. (Turkheimer, 'Three laws of behavior genetics and what they mean', p. 161. Bateson and Mameli, 'The innate and the acquired'.)

28 **Dynamic development was one of the big ideas of the twentieth century, and remains so.**

—⊸⊱ Without an infectious symbol like $E = mc^2$ or a phrase like 'nature versus nurture', this idea has been difficult to introduce to the public; few even bothered to try. Several decades passed while this transformative idea languished in obscurity and was eclipsed by other, more enthralling genetic headlines about Dolly the sheep, the Human Genome Project, 'criminal genes', and so on.

It languishes still. Meanwhile, in classrooms and baby nurseries everywhere, the oppressive reign of the gene-gift paradigm continues.

CHAPTER 2:
INTELLIGENCE IS A PROCESS, NOT A THING

PRIMARY SOURCES

American Psychological Association. 'Intelligence: Knowns and Unknowns. Report of a Task Force Established by the Board of Scientific Affairs of the American Psychological Association'. Released August 7, 1995.

Ceci, S. J. *On Intelligence: A Bio-ecological Treatise on Intellectual Development.* Harvard University Press, 1996.

Cravens, H. 'A scientific project locked in time: the Terman Genetic Studies of Genius'. *American Psychologist* 47, no. 2 (February 1992): 183–89.

Dickens, William T., and James R. Flynn. 'Heritability estimates versus large environmental effects: the IQ paradox resolved'. *Psychological Review* 108, no. 2 (2001): 346–69.

Dodge, Kenneth A. 'The nature-nurture debate and public policy'. *Merrill-Palmer Quarterly* 50, no. 4 (2004): 418–27.

Flynn, J. R. 'Beyond the Flynn Effect: Solution to All Outstanding Problems Except Enhancing Wisdom'. Lecture at the Psychometrics Centre, Cambridge Assessment Group, University of Cambridge, December 16, 2006.

Locurto, Charles. *Sense and Nonsense about IQ*. Praeger, 1991.

Risley, Todd R., and Betty Hart. *Meaningful Differences in the Everyday Experience of Young American Children*. Paul H. Brookes Publishing, 1995.

Schönemann, Peter H. 'On models and muddles of heritability'. *Genetica* 99, no. 2/3 (March 1997): 97–108.

Sternberg, Robert J. 'Intelligence, Competence, and Expertise'. In *Handbook of Competence and Motivation*, edited by A. J. Elliot and C. S. Dweck. Guilford Publications, 2005.

Sternberg, Robert J., and Janet E. Davidson. *Conceptions of Giftedness*. 1st ed. Cambridge University Press, 1986.

Sternberg, Robert J., and Elena Grigorenko. 'The predictive value of IQ'. *Merrill-Palmer Quarterly* 47, no. 1 (2001): 1–41.

CHAPTER NOTES

29 **[Some] assert that an individual's intelligence is a fixed quantity.**

Longer version: '[Some] assert that an individual's intelligence is a fixed quantity which cannot be increased. We must protest and react against this brutal pessimism . . . With practice, training, and above all method, we manage to increase our attention, our memory, our judgement, and literally to become more intelligent than we were before.' (Binet, *Les idées modernes sur les enfants*, pp. 105–6; this work has been reprinted in Elliot and Dweck, eds., *Handbook of Competence and Motivation*; see p. 124.)

30 **The good news is that, once learned, The Knowledge becomes literally embedded in the taxi driver's brain.**

Eleanor Maguire writes:

> Our finding that the posterior hippocampus increases in volume when there is occupational dependence on spatial navigation is evidence for functional differentiation within the hippocampus. In humans, as in other animals, the posterior hippocampus seems to be preferentially involved when previously learned spatial information is used, whereas

the anterior hippocampal region may be more involved (in combination with the posterior hippocampus) during the encoding of new environmental layouts.

A basic spatial representation of London is established in the taxi drivers by the time The Knowledge is complete. This representation of the city is much more extensive in taxi drivers than in the control subjects. Among the taxi drivers, there is, over time and with experience, a further fine-tuning of the spatial representation of London, permitting increasing understanding of how routes and places relate to each other. Our results suggest that the 'mental map' of the city is stored in the posterior hippocampus and is accommodated by an increase in tissue volume. (Maguire et al., 'Navigation-related structural change in the hippocampi of taxi drivers', pp. 4398–403.)

30 Further, her conclusion was perfectly consistent with what others have discovered in recent studies of violinists, Braille readers, meditation practitioners, and recovering stroke victims: that specific parts of the brain adapt and organise themselves in response to specific experience.

Leon Eisenberg surveys the evidence:

Colleagues . . . compared magnetoencephalographic recordings from experienced violinists with those from nonmusicians and found a substantially larger cortical representation of the fingers of the left hand (the one used to play the strings) than of the fingers of the right (or bowing) arm and more brain area dedicated to representation of fingers in the musicians than in the corresponding recordings from the nonmusicians.

A second example . . . is that the planum temporale is larger on the left than on the right in the musicians; the asymmetry is most marked in those with perfect pitch.

[Another study] found a substantial enlargement of hand representation in the three-finger Braille readers.

The cortex has a remarkable capacity for remodelling after environmental change. (Italics mine.) (Eisenberg, 'Nature, niche, and nurture', pp. 213–22.)

Eisenberg's citations:

Schlaug G., L. Jancke, Y. Huang, et al. 'Asymmetry in musicians'. *Science* 267 (1995): 699–701.

Elbert, Thomas, Christo Pantev, Christian Wienbruch, Brigitte Rockstroh, and Edward Taub. 'Increased cortical representation of the fingers of the left hand in string players'. *Science* 270 (1995): 305–7.

Sterr, A., M. M. Muller, T. Elbert, et al. 'Changed perceptions in Braille readers'. *Nature* 391 (1998): 134–35.

Yang, T. T., C. C. Gallen, and B. Schwartz. 'Sensory maps in the human brain'. *Nature* 368 (1994): 592–93.

Yang T. T., C. C. Gallen, V. S. Ramachandran, et al. 'Noninvasive detection of cerebral plasticity in adult human somatosensory cortex'. *Neuroreport* 5 (1994): 701–4.

Ramachandran, V. S., D. Rogers-Ramachandran, and M. Stewart. 'Perceptual correlates of massive cortical reorganization'. *Science* 258 (1992): 1159–60.

Ramachandran, V. S. 'Behavioral and magnetoencephalographic correlates of plasticity in the adult human brain'. *Proceedings of the National Academy of Sciences* 90 (1993): 10413–20.

Mogilner A., J. A. I. Grossman, and V. Ribary. 'Somatosensory cortical plasticity in adult humans revealed by magnetoencephalography'. *Proceedings of the National Academy of Sciences* 90 (1993): 3593–97.

30 **This is our famous 'plasticity': every human brain's built-in capacity to become, over time, what we demand of it.**

→ There are, of course, strict limits to plasticity. Every functioning human brain has an intricate and unchanging design, billions of years in the making. Various lobes and neural pathways are dedicated to specific functions: language, sensory input, consciousness, logical thought, abstract thought, spatial representation, and so on. The mind is not a blank slate. But this evolved design also includes an enormous capacity to learn and adapt, to hold specialised knowledge and wield specialised skills.

31 **Psychological methods of measuring intelligence:** Terman, *Genetic Studies of Genius*, vol. 1, p. v.

31 **Terman was part of a well-established movement convinced that intelligence was an inborn asset, inherited through genes, fixed at birth, and stable throughout life.**

→ Terman's direct mentor was the prominent psychologist (and first president of the American Psychological Association) G. Stanley Hall. H. Cravens writes:

> From his mentor, [G. Stanley] Hall, Terman learned that biological inheritance was all-powerful in determining the psyches and actions of animals and men . . . Hall's genetic psychology was a grand vision; simply put, Hall taught that minds have evolved through definite stages or types, from those of the lowliest cockroach to those of comparatively intellectual mammals and, finally, to those of the lower races, of children, of women, and then of rational White men. Hallian genetic psychology offered an overall hypothesis for Terman during his scientific career. (Cravens, 'A Scientific Project Locked in Time'.)

31 After Darwin published *On the Origin of Species* in 1859, Galton immediately sought to further define natural selection: Galton, *Hereditary Genius*, p. 2.

Galton also wrote:

> Biographies show [eminent men] to be haunted and driven by an incessant instinctive craving for intellectual work. They do not work for the sake of eminence, but to satisfy a *natural craving* for brain work, just as athletes cannot endure repose on account of their muscular irritability, which insists upon exercise. It is very unlikely that any conjunction of circumstances should supply a stimulus to brain work commensurate with what these men carry in their own constitutions. (Galton, *Hereditary Genius*, p. 80.)

31 In 1869, he published *Hereditary Genius,* arguing that smart, successful people were simply 'gifted' with a superior biology: Galton, *Hereditary Genius*, p. 39.

'The range of mental power between the greatest and least of English intellects is enormous,' Galton wrote. 'There is a continuity of natural ability reaching from one knows not what height, and descending to one can hardly say what depth.' (Galton, *Hereditary Genius*, p. 26.)

31 In 1874, he introduced the phrase 'nature and nurture' (as a rhetorical device to favour nature).

'The phrase "nature and nurture" is a convenient jingle of words,' Galton wrote, 'for it separates under two distinct heads the innumerable elements of which personality is composed. Nature is all that a man brings with himself into the world. Nurture is every influence from without that affects him after his birth.' (Galton, *English Men of Science*, p. 112.)

Galton probably got the phrase from Shakespeare's *The Tempest.*

Prospero: A devil, a born devil, on whose nature Nurture can never stick.

Judith Rich Harris suggests that Shakespeare may have got it from British writer Richard Mulcaster, who, thirty years earlier, had written, 'Nature makes the boy toward, nurture sees him forward.' (Harris, *The Nurture Assumption*, p. 4.)

31 In 1883, he invented 'eugenics', his plan to maximise the breeding of biologically superior humans and minimise the breeding of biologically inferior humans.

—⊸ Galton was an epic figure in the history of science. In his *New Yorker* review of Martin Brookes's recent Galton biography, Jim Holt eloquently explains his importance in two fields: eugenics and statistics.

Jim Holt on Galton's eugenics:

> In his long career, Galton didn't come close to proving the central axiom of eugenics: that, when it comes to talent and virtue, nature dominates nurture. Yet he never doubted its truth, and many scientists came to share his conviction. Darwin himself, in 'The Descent of Man', wrote, 'We now know, through the admirable labours of Mr. Galton, that genius . . . tends to be inherited.' Given this axiom, there are two ways of putting eugenics into practice: 'positive' eugenics, which means getting superior people to breed more; and 'negative' eugenics, which means getting inferior ones to breed less. For the most part, Galton was a positive eugenicist. He stressed the importance of early marriage and high fertility among the genetic elite, fantasising about lavish state-funded weddings in Westminster Abbey with the Queen giving away the bride as an incentive. Always hostile to religion, he railed against the Catholic Church for imposing celibacy on some of its most gifted representatives over the centuries. He hoped that spreading the insights of eugenics would make the gifted aware of their responsibility to procreate for the good of the human race. But Galton did not believe that eugenics could be entirely an affair of moral suasion. Worried by evidence that the poor in industrial Britain were breeding disproportionately, he urged that charity be redirected from them and toward the 'desirables'. To prevent 'the free propagation of the stock of those who are seriously afflicted by lunacy, feeble-mindedness, habitual criminality, and pauperism,' he urged 'stern compulsion', which might take the form of marriage restrictions or even sterilisation.
>
> Galton's proposals were benign compared with those of famous

contemporaries who rallied to his cause. H. G. Wells, for instance, declared, 'It is in the sterilisation of failures, and not in the selection of successes for breeding, that the possibility of an improvement of the human stock lies.' Although Galton was a conservative, his creed caught on with progressive figures like Harold Laski, John Maynard Keynes, George Bernard Shaw, and Sidney and Beatrice Webb. In the United States, New York disciples founded the Galton Society, which met regularly at the American Museum of Natural History, and popularisers helped the rest of the country become eugenics-minded. 'How long are we Americans to be so careful for the pedigree of our pigs and chickens and cattle – and then leave the ancestry of our children to chance or to "blind" sentiment?' asked a placard at an exposition in Philadelphia. Four years before Galton's death, the Indiana legislature passed the first state sterilisation law, 'to prevent the procreation of confirmed criminals, idiots, imbeciles, and rapists.' Most of the other states soon followed. In all, there were some sixty thousand court-ordered eugenically unfit. It was in Germany that eugenics took its most horrific form. Galton's creed had aimed at the uplift of humanity as a whole; although he shared the prejudices that were common in the Victorian era, the concept of race did not play much of a role in his theorising.

German eugenics, by contrast, quickly morphed into *Rassenhygiene* – race hygiene. Under Hitler, nearly four hundred thousand people with putatively hereditary conditions like feeblemindedness, alcoholism, and schizophrenia were forcibly sterilised. In time, many were simply murdered. The Nazi experiment provoked a revulsion against eugenics that effectively ended the movement. (Holt, 'Measure for Measure', p. 90.)

Jim Holt on Galton's statistical inventions:

After obtaining height data from two hundred and five pairs of parents and nine hundred and twenty-eight of their adult children, Galton plotted the points on a graph, with the parents' heights represented on one axis and the children's on the other. He then pencilled a straight line though the cloud of points to capture the trend it represented. The slope of this line turned out to be two-thirds. What this meant was that exceptionally tall (or short) parents had children who, on average, were only two-thirds as exceptional as they were. In other words, when it came to height children tended to be less exceptional than their parents. The same, he had noticed years earlier, seemed to be true in the case of 'eminence': the children of J. S. Bach, for example, may have been more musically distinguished than average,

> but they were less distinguished than their father. Galton called this phenomenon 'regression toward mediocrity'.
>
> Regression analysis furnished a way of predicting one thing (a child's height) from another (its parents') when the two things were fuzzily related. Galton went on to develop a measure of the strength of such fuzzy relationships, one that could be applied even when the things related were different in kind – like rainfall and crop yield. He called this more general technique 'correlation'. The result was a major conceptual breakthrough. Until then, science had pretty much been limited to deterministic laws of cause and effect – which are hard to find in the biological world, where multiple causes often blend together in a messy way. Thanks to Galton, statistical laws gained respectability in science. His discovery of regression toward mediocrity – or regression to the mean, as it is now called – has resonated even more widely. (Holt, 'Measure for Measure', pp. 88–89.)

32 **'[the word] "intelligence" has become a mere vocal sound':** Spearman, *The Abilities of Man, Their Nature and Measurement*, cited in Schönemann, 'On models and muddles of heritability'.

This was still the case in the 1980s. From the American Psychiatric Association report: 'Indeed, when two dozen prominent theorists were recently asked to define intelligence, they gave two dozen somewhat different definitions.' (Hertzig and Farber, eds., *Annual Progress in Child Psychiatry and Child Development 1997*, p. 96.)

32 **there must be a single 'general intelligence' (*g* for short):** Spearman, 'General intelligence, objectively determined and measured', pp. 201–93; Green, Classics in the History of Psychology Web site.

32 **'*G* is, in the normal course of events, determined innately,' Spearman declared. 'A person can no more be trained to have it in higher degree than he can be trained to be taller':** Deary, Lawn, and Bartholomew, 'A conversation between Charles Spearman, Godfrey Thomson, and Edward L. Thorndike', p. 128.

→ In the absence of any persuasive alternative, Spearman's *g* resonated with the psychological community and proved quite resilient throughout the twentieth century. His *g* was further refined in the 1970s and '80s by Berkeley psychologist Arthur Jensen and gained considerable traction in the psychological community.

That's not to say that Jensen won over a clear majority of academic psychologists. But he clearly won over at least a large plurality. 'Of the 60 papers in our sample, 29 cited Jensen's article negatively. This number includes articles that took exception to almost every point presented in the paper. It also includes those in which the authors debated specific points Jensen made. Eight of the articles cited Jensen's paper as an example of a controversy. Eight more used the article as a background reference. Only fifteen of the articles cited Jensen in agreement with his positions, and seven of them only on minor points. Further readings have confirmed that our sample is typical of the way authors have cited the Jensen work.' ('High Impact Science and the Case of Arthur Jensen', pp. 652–62.)

In 1971, Raymond Cattell divided *g* into two independent subcomponents – fluid intelligence (*g*F) and crystallised intelligence (*g*C). Fluid intelligence was thought to be a fixed, innate ability to reason and conceptualise; crystallised intelligence was the school-influenced ability to draw on knowledge and experience.

Throughout the twentieth century, psychologists supporting general intelligence became naturally allied with the psychologists supporting 'heritability' from twin studies, and together they painted a formidable neo-Galtonian portrait of humans with preset genetic capabilities. Collectively, these modern Galton disciples became known as 'behaviour geneticists'. In the 1980s and '90s, they published a slew of studies aiming to solidify their position and influence policy. In short, they wanted to steer resources toward the innately superior and not waste much on the genetically inferior.

Kenneth A. Dodge writes: 'The naïve hope that early environmentalists could be easily manipulated to alter long-term outcomes inspired a backlash of behaviour-genetic studies in the 1980s and 1990s that championed the high per cent of variance in behaviour that is accounted for by genes. The legacy of this backlash is the argument that public and private resources (e.g., the best schools and highest incomes) should be administered according to the selection of those with the highest (presumably, genetically based) potential to achieve, rather than to compensate for biological or environmental disadvantage. The scholarly anchor or the policy conclusion was exemplified in the essays by Scarr (1992), Lytton (1990), and Harris (1995, 1998) which claimed that the environment accounts for very little influence on human behaviour. After 50 years of study, it seemed that little had been learned.' (Dodge, 'The nature-nurture debate and public policy', pp. 418–27.)

32 In 1916, Stanford's Lewis Terman produced a practical equivalent of *g* with his Stanford-Binet Intelligence Scales.

Excerpt from an excellent article by Mitchell Leslie:

In 1916, Terman sprang his test on America. He released *The Measurement of Intelligence*, a book that was half instruction manual and IQ test, half manifesto for universal testing. His little exam, which a child could complete in a mere 50 minutes, was about to revolutionise what students learned and how they thought of themselves.

Few American children have passed through the school system in the last 80 years without taking the Stanford-Binet or one of its competitors. Terman's test gave U.S. educators the first simple, quick, cheap and seemingly objective way to 'track' students, or assign them to different course sequences according to their ability. The following year, when the United States entered World War I, Terman helped design tests to screen Army recruits. More than 1.7 million draftees took his tests, broadening public acceptance of widespread IQ testing.

The Stanford-Binet made Terman a leader in a fervent movement to take testing far beyond the schoolhouse and Army base. Proponents considered intelligence the most valuable human quality and wanted to test every child and adult to determine their place in society. The 'intelligence-testers' – a group that included many eugenicists – saw this as the tool for engineering a fairer, safer, fitter and more efficient nation, a 'meritocracy' run by those most qualified to lead. In their vision of a vibrant new America, IQ scores would dictate not only what kind of education a person received but what work he or she could get. The most important and rewarding jobs in business, the professions, academia and government would go to the brightest citizens. People with very low scores – under about 75 – would be institutionalised and discouraged or prevented from having children.

IQ tests and the social agenda of their advocates roused critics right from the start. To the journalist Walter Lippmann, the intelligence-testers were 'the Psychological Battalion of Death', seizing unparalleled power over every child's future. Lippmann and Terman duelled in the pages of *The New Republic* in 1922 and 1923. 'I hate the impudence of a claim that in 50 minutes you can judge and classify a human being's predestined fitness in life,' Lippmann wrote. 'I hate the sense of superiority which it creates, and the sense of inferiority which it imposes.' In a sarcastic rejoinder, Terman compared Lippmann to the creationist William Jennings Bryan and other opponents of scientific progress, then attacked Lippmann's writing style as 'much too verbose for literal quotation.' Though he could never match Lippmann's eloquence, in the end Terman won the war: intelligence testing continued to spread. By the

> 1930s, kids with high IQs were being sent into more challenging classes to prepare for high-earning jobs or college, while low scorers got less demanding coursework, reduced expectations and dimmer job prospects. (Leslie, 'The Vexing Legacy of Lewis Terman'.)

32 **adapted from an earlier version by French psychologist Alfred Binet.**

—∘⋟ Ironically, IQ tests were not originally intended to measure a person's intelligence at all. First invented in 1905 by psychologist Alfred Binet and physician Theodore Simon as an effort to identify French schoolchildren in need of most attention, the Binet-Simon test aimed to lift students up rather than assign them a permanent intellectual rank.

'The procedures which I have indicated will, if perfected, come to *classify* a person before or after such another person or such another series of persons,' wrote Binet. 'But I do not believe that one may measure one of the intellectual *aptitudes* in the sense that one measures length or a capacity' (italics mine). (Varon, 'Alfred Binet's concept of intelligence', p. 41.)

'With practice, training, and above all method,' Binet wrote in 1909, 'we manage to increase our attention, our memory, our judgment, and literally to become more intelligent than we were before.' (A century later, the science of motivation and expert performance would validate this.) (Binet, *Les idées modernes sur les enfants*, pp. 105–6; this work has been reprinted in Elliot and Dweck, eds., *Handbook of Competence and Motivation*; see p. 124.)

Mitchell Leslie adds:

> With questions ranging from mathematical problems to vocabulary items, the Americanised test was supposed to capture 'general intelligence', an innate mental capability that Terman felt was as measurable as height and weight. As a hardcore hereditarian, he believed that genetics alone dictated one's level of general intelligence. This vital constant, which he called an 'original endowment', wasn't altered by education or home environment or hard work, he maintained. To denote it, he selected the term 'intelligence quotient'. (Leslie, 'The Vexing Legacy of Lewis Terman'.)

33 **the National Intelligence Test (a precursor to the SAT) was designed by Edward Lee Thorndike:** Saretzky, 'Carl Campbell Brigham, the Native Intelligence Hypothesis, and the Scholastic Aptitude Test'.

33 **Princeton psychologist Carl Brigham disavowed his own creation, writing that all intelligence tests were based on 'one of the most glorious**

fallacies in the history of science, namely that the tests measured native intelligence purely and simply without regard to training or schooling.'

Matt Pacenza writes:

> In an unpublished manuscript which Lemann unearthed, Brigham wrote that the standardised testing movement was based on 'one of the most glorious fallacies in the history of science, namely that the tests measured native intelligence purely and simply without regard to training or schooling. The test scores very definitely are a composite including schooling, family background, familiarity with English and everything else. (Pacenza, 'Flawed from the Start'; Lemann, *The Big Test.*)

34 **diagram illustrating Distribution of IQ Scores:** Locurto, *Sense and Nonsense About IQ*, p. 5.

—•» As Stephen Jay Gould outlines, Terman assigned a protégée, Catherine Cox, to look back in time and assign IQs to dead geniuses – a logical farce considering what the IQ is supposed to do. They assigned a score of 200 to Terman's hero Galton. (Gould, *The Mismeasure of Man*, pp. 213–17.)

At the time it was introduced, Terman's test filled a particular need in American schools and society. In that age of standardisation and mechanisation, American culture was obsessed with establishing consistent measures in all walks of life. IQ scores provided an easy way to separate the most promising students from the least promising, to identify and nurture future leaders in business, government, the military, and so on. 'Tests of "general intelligence", given as early as six, eight, or ten years,' Terman insisted with pride, 'tell a great deal about the ability to achieve either presently or 30 years hence.'

Terman was correct to suggest a strong connection between academic skills and success in modern, industrialised society. Someone who performs well in school and in abstract intellectual tests is generally (albeit with many obvious exceptions) more likely to succeed in business, law, journalism, and of course academia – any profession that puts a premium on any of those same skills. For that reason, IQ scores have proven to be generally predictive of success in Western societies where success is sufficiently based on education.

Sternberg and Grigorenko add:

> IQ seems to be predictive of the reaching of all steps of career life in a stable society, where Western schooling is valued and rewarded, income is scaled in rough correspondence to years of education, and highly-skilled labor is needed. (Sternberg and Grigorenko, 'The predictive value of IQ', p. 9.)

33 **at its core, IQ was merely a population-sorting tool.**
Just as Binet had originally intended.

33–34 Lewis Terman and colleagues actually recommended that individuals identified as 'feebleminded' by his test be removed from society and that anyone scoring less than 100 be automatically disqualified from any prestigious position.

Bonnie Strickland writes:

> Terman (1916) actually appealed for universal intelligence testing, believing that the enormous costs of crime and vice could be reduced by removing the feebleminded from society. Further, theorising that employment opportunities should be determined by intelligence, Terman proposed a social order that would close prestigious and rewarding professions to people with IQs under 100. (Strickland, 'Misassumptions, misadventures, and the misuse of psychology', p. 333 – citing Terman, *The Intelligence of School Children*.)

—◦◈ The Terman book is fascinating reading. Although Terman's IQ test could not really prove either fixed or innate intelligence, he maintained that it had proved both and proceeded accordingly. Terman's logic was simple: since his tests showed a reasonable consistency over the years, they revealed that intelligence was innate and fixed. (Terman, *The Intelligence of School Children*.)

The French did not share this leave-them-behind approach, and to this day they largely ignore modern IQ tests. (Sternberg and Grigorenko, 'The predictive value of IQ', p. 2.)

34 **'does not imply unchangeability'**: Howe, 'Can IQ Change?', p. 71.

34–35 'IQ scores,' explains Cornell University's Stephen Ceci, 'can change quite dramatically as a result of changes in family environment (Clarke, 1976; Svendsen, 1982), work environment (Kohn, 1981), historical environment (Flynn, 1987), styles of parenting (Baumrind,

1967; Dornbusch, 1987), and, most especially, shifts in level of schooling': Ceci, *On Intelligence*, p. 73.

Ceci's Citations

FAMILY ENVIRONMENT

Clarke, Ann M., and Alan D. Clarke. *Early Experience and the Life Path.* Somerset, 1976.

Svendsen, Dagmund. 'Factors related to changes in IQ: a follow-up study of former slow learners'. *Journal of Child Psychology and Psychiatry* 24, no. 3 (1983): 405–13.

WORK ENVIRONMENT

Kohn, Melvin, and Carmi Schooler. 'The Reciprocal Effects of the Substantive Complexity of Work and Intellectual Flexibility: A Longitudinal Assessment'. *American Journal of Sociology* 84 (July 1978): 24–52.

HISTORICAL ENVIRONMENT

Flynn, J. R. 'Massive IQ gains in 14 nations: what IQ tests really measure'. *Psychological Bulletin* 101 (1987): 171–91.

STYLES OF PARENTING

Baumrind, D. 'Child care practices anteceding three patterns of preschool behavior'. *Genetic Psychology Monographs* 75 (1967): 43–88.

Dornbusch, Sanford M., Philip L. Ritter, P. Herbert Leiderman, Donald F. Roberts, and Michael J. Fraleigh. 'The relation of parenting style to adolescent school performance'. *Child Development* 58, no. 5 (October 1987): 1244–57.

—•➾ Lewis Terman's most important claim for IQ – that it reveals a person's fixed, innate intelligence – relies entirely on the assertion that individual IQ scores remain the same throughout people's lives. *This simply is not true.* While one study reported a majority of people's scores changing relatively little over time, that same study reported that, 'in a nontrivial minority of children, naturalistic IQ change is marked and real.' Other large studies showed a significant majority of students experiencing an IQ swing of 15 points or more over time. (Sternberg and Grigorenko, 'The predictive value of IQ', p. 13.)

It also means that Spearman's IQ test has ironically sowed the seeds of its own destruction. In so efficiently documenting narrow bands of academic achievement decade after decade, the test that he devised to prove the fixedness of intelligence inadvertently demonstrated how flexible and buildable intelligence really is.

James Flynn: 'At any particular time, factor analysis will extract *g*(iQ) – and intelligence appears unitary. Over time, real-world cognitive skills assert their functional autonomy and swim freely of *g* – and intelligence appears multiple. If you want to see *g*, stop the film and extract a snap shot; you will not see it while the film is running. Society does not do factor analysis.' (Flynn, *What Is Intelligence?* p. 18.)

IQ is as changeable as much as 30 points, as reported in Sherman and Key; and as much as 18 points, as reported in Jones and Bayley. (Sherman and Key results reported in Ceci, *On Intelligence*, chapter 5; Jones and Bayley, 'The Berkeley Growth Study', pp. 167–73.)

35 **Their unavoidable conclusion was that 'children develop only as the environment demands development':** Sherman and Key, 'The intelligence of isolated mountain children', pp. 279–90.

—⊸ Other studies have demonstrated that IQ scores drift lower during the summer months (except for those attending an academic camp) and that they rise steadily as the school year progresses. In other words, schooling itself has a direct effect on IQ scores. 'Contrary to the traditional belief that information contained on IQ tests is potentially available to all children, regardless of environmental conditions,' writes Stephen Ceci, 'it has been known for many decades that a child's experience of schooling exerts a strong influence on intelligence test performance . . . This relationship is still substantial after potentially confounding variables, such as the tendency for the most intelligent children to begin schooling earlier and remain there longer, are controlled.' (Ceci, *On Intelligence*, chapter 5.)

To the extent that scores did show some stability across a large population, it seemed largely a function not of innate intelligence but population inertia. Inertia is the tendency for things to remain in their same relative state – of rest or motion – unless and until something comes along to change the dynamic. It's true of molecular physics and it's equally true of human action and populations. Most people performing at the middle of the intellectual pack at age ten are going to be performing at the middle of the intellectual pack at age twenty or thirty. This observation says nothing about intelligence; it's simple population dynamics. You could say the same thing about almost any trait: by and large, the funniest ten-year-olds are also going to be the funniest twenty-year-olds, the fastest ten-year-olds are also going to be the fastest twenty-year-olds; the biggest-toed ten-year-olds are also going to be the biggest-toed twenty-year-olds. There will be plenty of individual exceptions, but in a large group, this consistency of order is always going to be the norm.

Another way of illustrating population inertia is to consider the annual New York City marathon, with its ninety thousand runners. If one were to list the order of runners at the ten-mile mark, and then compare that order to the order at the finish line, you would find a very solid correlation. Almost none of the runners at the finish would be in exactly the same position as before, and of course some would be way off, but on the whole, the correlation of runners' ten-mile positions to twenty-six-mile positions would be very high. Why? Because by mile ten, runners have already established their pace, their level of endurance, their level of competitiveness, and so on; the pack has taken shape and will keep roughly the same shape throughout the race. Obviously, this correlation has absolutely nothing to do with the underlying cause of each runner's performance. It simply reflects the dynamic of any competition.

So it is with IQ. Without question, there are wide differences in intellectual abilities throughout life, and if you test one hundred thousand kids at age ten and then test them again at age twenty-six, you're going to find that, on average, they remain in roughly the same intellectual pecking order. Many individual scores will diverge – IQ scores are known to swing as much as thirty points over time in individuals with changing circumstances – but as a group, the age-ten numbers will correlate rather well with the age-twenty-six numbers.

Surprise, surprise: most people who are pretty good at academics at age ten (compared to others the same age) are also pretty good at age twenty-six; most who are excellent at age ten are also excellent at age twenty-six. That's what IQ stability tells us – and that's *all* it tells us. It does not suggest inborn limits, and it doesn't even hint at the extraordinary power of individuals to change their own circumstances and lift their intellectual performance.

Intelligence scores of infants are *not* predictive of future scores or life success. That population is still too much in flux; individuals have not yet hit their stride; the pack has not yet taken shape; population inertia has not yet set in.

35 **Comparing raw IQ scores over nearly a century, Flynn saw that they kept going up:** Nippert, 'Eureka!'

35 **IQ test takers improved over their predecessors by three points every ten years.**

—⊸ These comparisons draw on the raw scores – not the weighted scores that are annually recalibrated so that the average is always 100.

35 Using a late-twentieth-century average score of 100, the comparative score for the year 1900 was calculated to be about 60 – leading to the truly absurd conclusion, acknowledged Flynn, 'that a majority of our ancestors were mentally retarded.'

—⊸ This retroactive analysis illustrates the logical flaw in continually using a curved IQ score to dismiss the competence of anyone scoring below 100.

36 '[The intelligence of] our ancestors in 1900 was anchored in everyday reality,' explains Flynn. 'We differ from them in that we can use abstractions and logic and the hypothetical.'

Flynn adds:

> When [asked]: 'What do dogs and rabbits have in common,' Americans in 1900 would be likely to say, 'You use dogs to hunt rabbits.' The correct [contemporary test] answer, that both are mammals, assumes that the important thing about the world is to classify it in terms of the taxonic categories of science . . . Our ancestors found pre-scientific spectacles more comfortable than post-scientific spectacles, [because that's what] showed them what they considered to be most important about the world . . . (Flynn, 'Beyond the Flynn Effect'.)

36 Examples of abstract notions that simply didn't exist in the minds of our nineteenth-century ancestors include the theory of natural selection (formulated in 1864), and the concepts of control group (1875) and random sample (1877).

This comes from a 2006 lecture by James Flynn. An extended excerpt:

> Over the last century and a half, science and philosophy have expanded the language of educated people, particularly those with a university education, by giving them words and phrases that greatly increase their critical acumen. Each of these terms stands for a cluster of interrelated ideas that virtually spell out a method of critical analysis applicable to social and moral issues. I will call them *'shorthand abstractions'* (or SHAs), it being understood that they are abstractions with peculiar analytic significance.

I will name [some] SHAs followed by the date they entered educated usage (dates all from the Oxford English Dictionary on line):

(1) Market (1776: economics). With Adam Smith, this term altered from the merely concrete (a place where you bought something) to an abstraction (the law of supply and demand). It provokes a deeper analysis of innumerable issues. If the government makes university education free, it will have to budget for more takers. If you pass a minimum wage, employers will replace unskilled workers with machines, which will favour the skilled. If you fix urban rentals below the market price, you will have a shortage of landlords providing rental properties. Just in case you think I have revealed my politics, I think the last a strong argument for state housing.

(2) Percentage (1860: mathematics). It seems incredible that this important SHA made its debut into educated usage less than 150 years ago. Its range is almost infinite. Recently in New Zealand, there was a debate over the introduction of a contraceptive drug that kills some women. It was pointed out that the extra fatalities from the drug amounted to 50 in one million (or 0.005%) while without it, an extra 1000 women (or 0.100%) would have fatal abortions or die in childbirth.

(3) Natural selection (1864: biology). This SHA has revolutionised our understanding of the world and our place in it. It has taken the debate about the relative influences of nature and nurture on human behaviour out of the realm of speculation and turned it into a science. Whether it can do anything but mischief if transplanted into the social sciences is debatable. It certainly did harm in the 19th century when it was used to develop foolish analogies between biology and society. Rockefeller was acclaimed as the highest form of human being that evolution had produced, a use denounced even by William Graham Sumner, the great 'Social Darwinist'. I feel it made me more aware that social groups superficially the same were really quite different because of their origins. Black unwed mothers who are forced into that status by the dearth of promising male partners are very different from unwed mothers who choose that status because they genuinely prefer it.

(4) Control group (1875: social science). Recognition that before and after comparisons of how interventions affect people are usually flawed. We introduce an enrichment programme in which pre-school children go to a 'play centre' each day. It is designed to raise the IQ of children at risk of being diagnosed as mentally retarded. Throughout the programme we test their IQs to monitor progress. The question arises, what has raised their IQs? The enrichment programme, getting out of a dysfunctional home for 6 hours each day, the lunch they had at

the play centre, the continual exposure to IQ tests. Only a control group selected from the same population and subjected to everything but the enrichment programme can suggest an answer.

(5) Random sample (1877: social science). Today, the educated public is much more likely to spot biased sampling than they were a few generations ago. In 1936, the Literary Digest telephone poll showed that Landon was going to beat Roosevelt for President and was widely believed, even though few had telephones except the more affluent.

(6) Naturalistic fallacy (1903: moral philosophy). That one should be wary of arguments from facts to values, for example, an argument that because something is a trend in evolution it provides a worthy goal for human endeavour.

(7) Charisma effect (1922: social science). Recognition that when a technique is applied by a charismatic innovator or disciples fired by zeal, it may be successful for precisely that reason. For example, a new method of teaching mathematics often works until it is used by the mass of teachers for whom it is merely a new thing to try.

(8) Placebo (1938: medicine). The recognition that merely being given something apparently endorsed by authority will often have a salutary effect for obvious psychological reasons. Without this notion, a rational drugs policy would be overwhelmed by the desperate desire for a cure by those stricken with illness.

(9) Falsifiable/tautology (1959: philosophy of science). The stipulation that a factual claim is bankrupt (a mere tautology or closed circle of definitions) unless it is testable against evidence. It can be used to explode: a theory of motivation that asserts all human acts are selfish and yet rules out every possible counterexample; the claim that 'real' workers by definition have a revolutionary psychology; that 'real' Christians are always charitable; and so forth. (Flynn, 'Beyond the Flynn Effect'.)

Flynn and his colleague William Dickens add:

Thanks to industrialisation, it is likely that the cognitive complexity of the average person's job has increased over the last century. There is no doubt that more-demanding educational credentials control access to a wide range of jobs. There are far more people in scientific, managerial, and technical positions than ever before. Increased leisure time is another possible trigger for IQ gains, as some activities undertaken during extended leisure (reading, puzzles, games such as chess) may be honing people's facilities. Radio and television may be factors. It is possible that the machinery we increasingly surround ourselves with (e.g., cars, phones, computers, and VCRs) have increased the demands on our cognitive capacities. The shift to fewer children in each family, affording more time to cater to children's curiosity and richer individual interactions, may have played a role. Some or all of these may have contributed to a significant attitude shift: The current generation may take abstract problem solving far more seriously than preceding generations did. The direct effects of these changes need not be large. But because they are widespread and persistent trends, they could loom large relative to the many less-constant environmental influences that produce most differences between people. (Dickens and Flynn, 'Heritability estimates versus large environmental effects', pp. 346–69.)

37 **Perhaps the most striking of Flynn's observations is this: 98 per cent of IQ test takers today score better than the average test taker in 1900.**

Flynn writes:

> The Wechsler-Binet rate of gain (0.3 points per year) entails that the school children of 1900 would have had a mean IQ just under 70. The Raven-Similarities rate (0.5 points per year) yields a mean IQ of 50 (against current norms). (Flynn, 'Beyond the Flynn Effect'.)

—⊸ This is arguably the most important observation in this book.

37 **'Our ability to improve the academic accomplishment of students'**: Murray, 'Intelligence in the Classroom'.

37–38 **'Even the best schools under the best conditions cannot repeal the limits'**: Charles Murray, 'Intelligence in the Classroom'.

—⊸ *Is Charles Murray a straw man?* This question was raised by some draft readers of this book. *Aren't his views so ridiculous and outside the mainstream that they aren't worth critiquing?*

Actually, Murray's views on this subject command a good deal of respect and mainstream attention. He is a fellow at the widely respected American Enterprise Institute in Washington, D.C. He continues to write for the *Wall Street Journal*, the *New York Times*, and the *Weekly Standard* and appears on C-SPAN.

38 **'small to moderate'**: 'Head Start Impact Study, First Year Findings', June 2005, Prepared for Office of Planning, Research and Evaluation

Administration for Children and Families, U.S. Department of Health and Human Services, Washington, D.C., Westat, The Urban Institute Chesapeake Research Associates Decision Information Resources Inc., and American Institutes for Research.

38 **Children in professionals' homes were exposed to an average of more than fifteen hundred more spoken words per hour than children in welfare homes.**

⟶ And more than three times the children of parents on welfare. Actual numbers: welfare children 616 words per hour; professionals' kids 2,153 words per hour. Estimate based on fourteen-hour day. Words were spoken live in person – not on TV or radio. (Hart and Risley, 'The early catastrophe'.)

39 **Not surprisingly, the psychological community responded with a mixture of interest and deep caution. In 1995, an American Psychological Association task force wrote that 'such correlations may be mediated by genetic as well as (or instead of) environmental factors.' Note 'instead of'. In 1995, it was still possible for leading research psychologists to imagine that better-off kids could be simply inheriting smarter genes from smarter parents, that spoken words could be merely a genetic effect and not a cause of anything.**

From the APA report:

> There is no doubt that such variables as resources of the home and parents' use of language are correlated with children's IQ scores, but such correlations may be mediated by genetic as well as (or instead of) environmental factors. Behaviour geneticists frame such issues in quantitative terms. As noted in Section 3, environmental factors certainly contribute to the overall variance of psychometric intelligence. But how much of that variance results from differences between families, as contrasted with the varying experiences of different children in the same family? Between-family differences create what is called 'shared variance' or c2 (all children in a family share the same home and the same parents). Recent twin and adoption studies suggest that while the value of c2 (for IQ scores) is substantial in early childhood, it becomes quite small by late adolescence. These findings suggest that differences in the life styles of families, whatever their importance may be for many aspects of children's lives, make little long-term difference for the skills measured by intelligence tests. We should note, however, that low-income and non-white families are poorly represented in existing adoption studies as well as in most twin samples. Thus it is not yet clear whether these

> surprisingly small values of (adolescent) c2 apply to the population as a whole. It remains possible that, across the full range of income and ethnicity, between-family differences have more lasting consequences for psychometric intelligence. (APA, 'Stalking the Wild Taboo'.)

39 Now we know better. We know that genetic factors do not operate 'instead of' environmental factors, they interact with them: GxE.

—⊸ Recall Massimo Pigliucci's observation: 'Biologists have come to realise that if one changes *either* the genes *or* the environment, the resulting behaviour can be dramatically different. The trick, then, is not in partitioning causes between nature and nurture, but in [examining] the way genes and environments interact dialectically to generate an organism's appearance and behaviour.' (Pigliucci, 'Beyond nature and nurture', pp. 20–22.)

39 Speaking to children early and often. This trigger was revealed in Hart and Risley's incontrovertible study and reinforced by the University of North Carolina's Abecedarian Project, which provided environmental enrichment to children from birth, with the study subjects showing substantial gains compared with a control group.

—⊸ For example, in the North Carolina 'Abecedarian Project' – an all-day programme that provided various forms of environmental enrichment to fifty-seven children from infancy onward (mean starting age 4.4 months) and compared their test performance to a matched control group – differences between groups became apparent before the end of the first year. The difference did not diminish over time; the IQ difference between the groups was still present at age twelve. (Neisser, 'Rising Scores on Intelligence Tests'.)

39–40 Reading early and often. In 2003, a national study reported the positive influence of early parent-to-child reading, regardless of parental education level. In 2006, a similar study again found the same thing about reading, this time ruling out any effects of race, ethnicity, class, gender, birth order, early education, maternal education, maternal verbal ability, and maternal warmth.

Helen Raikes and colleagues write:

> A national study of preschool-aged children participating in Head Start demonstrated that, compared with parents who read less frequently, more frequent reading in the fall was associated with both higher concurrent scores on literacy measures and larger gains during the year, even after controlling for parental education level, parental literacy level, and the presence of books in the home. Children of parents

who reported reading to them 'not at all' or 'only once or twice a week' had receptive vocabulary scores that were lower than those of children whose parents reported reading 'three to six times a week'. Reading three to six times per week was associated with greater fall-to-spring vocabulary gains than was reading less frequently, and children whose parents reported reading daily had even larger gains. In addition, some research suggests that earlier regular experience with bookreading, beginning as young as 14 months, is particularly beneficial.

In regression analyses to examine relations between reading and child outcomes, we controlled for the variables of race/ethnicity, demographic risk, maternal education and verbal ability, gender, birth order, Early Head Start enrolment, and maternal warmth. In the English-speaking group, at 14 months, reading several times weekly or reading daily was significantly related to vocabulary and comprehension. Findings were similar for vocabulary and MDI scores at 24 months, even after controlling for children's 14-month vocabulary. A pattern of daily reading over three data points significantly related to child language and cognitive outcomes at 36 months. Reading daily at a minimum of one of the periods predicted language outcomes for Spanish-speaking children. Regression path analyses showed paths from early reading to later reading, early vocabulary to later child language outcomes, and 14-month vocabulary to 24-month reading. Paths for concurrent reading revealed associations with vocabulary at 14 and 24 months. (Raikes et al., 'Mother-child bookreading in low-income families', pp. 940–43.)

40 Nurturance and encouragement. Hart and Risley also found that, in the first four years after birth, the average child from a professional family receives 560,000 more instances of encouraging feedback than discouraging feedback; a working-class child receives merely 100,000 more encouragements than discouragements; a welfare child receives 125,000 more discouragements than encouragements.

Hart and Risley write:

> But the children's language experience did not differ just in terms of the number and quality of words heard. We can extrapolate similarly the relative differences the data showed in children's hourly experience with parent affirmatives (encouraging words) and prohibitions. The average child in a professional family was accumulating 32 affirmatives and five prohibitions per hour, a ratio of 6 encouragements to 1 discouragement. The average child in a working-class family was accumulating 12 affirmatives and seven prohibitions per hour, a ratio of 2

encouragements to 1 discouragement. The average child in a welfare family, though, was accumulating five affirmatives and 11 prohibitions per hour, a ratio of 1 encouragement to 2 discouragements. In a 5,200-hour year, that would be 166,000 encouragements to 26,000 discouragements in a professional family, 62,000 encouragements to 36,000 discouragements in a working-class family, and 26,000 encouragements to 57,000 discouragements in a welfare family.

Extrapolated to the first four years of life, the average child in a professional family would have accumulated 560,000 more instances of encouraging feedback than discouraging feedback, and an average child in a working-class family would have accumulated 100,000 more encouragements than discouragements. But an average child in a welfare family would have accumulated 125,000 more instances of prohibitions than encouragements. By the age of 4, the average child in a welfare family might have had 144,000 *fewer* encouragements and 84,000 *more* discouragements of his or her behaviour than the average child in a working-class family. Extrapolating the relative differences in children's hourly experience allows us to estimate children's cumulative experience in the first four years of life and so glimpse the size of the problem facing intervention. Whatever the inaccuracy of our estimates, it is not by an order of magnitude such that 60,000 words becomes 6,000 or 600,000. Even if our estimates of children's experience are too high by half, the differences between children by age 4 in amounts of cumulative experience are so great that even the best of intervention programmes could only hope to keep the children in families on welfare from falling still further behind the children in the working-class families. (Hart and Risley, 'The early catastrophe'.)

40 **Setting high expectations.**

Studies validating this finding:

Edmonds, R. 'Characteristics of Effective Schools'. In *The School Achievement of Minority Children: New Perspectives*, edited by U. Neisser. Lawrence Erlbaum, 1986, pp. 93–104.

Rutter, M., B. Maughan, P. Mortimore, J. Ouston, and A. Smith. *Fifteen Thousand Hours*. Harvard University Press, 1979.

Slavin, R., N. Karweit, and N. Madden. *Effective Programs for Students at Risk*. Allyn and Bacon, 1989.

Ellen Winner: 'Parents of gifted children typically have high expectations, and also model hard work and high achievement themselves.' (Winner, 'The origins and ends of giftedness', pp. 159–69.)

Winner's Citations

Bloom, B. *Developing Talent in Young People*. Ballantine, 1985.

Csikszentmihályi, Mihály, Kevin Rathunde, and Samuel Whalen. *Talented Teenagers*. Cambridge University Press, 1993.

Gardner, H. *Creating Minds: An Anatomy of Creativity Seen Through the Lives of Freud, Einstein, Picasso, Stravinsky, Eliot, Graham, and Gandhi*. Basic Books, 1993.

40 **Embracing failure.**

'Deliberate practice does not involve a mere execution or repetition of already attained skills but repeated attempts to reach beyond one's current level which is associated with frequent failures.' (Ericsson et al., 'Giftedness and evidence for reproducibly superior performance', pp. 3–56.)

40 **Encouraging a 'growth mindset':** Dweck, *Mindset: The New Psychology of Success*.

41 **phenomenon that we might call 'carton calculus':** Ceci, *On Intelligence*, p. 33.

42 **Halfway around the world, in Kisumu, Kenya, Yale psychologist Robert Sternberg stumbled on exactly the same phenomenon in 2001 when studying the intelligence of Dholuo schoolchildren.**

—◦≫ Surprisingly, Sternberg found a 'significantly negative' correlation between his herbal medicine test and an English language test and no significant correlation between his test and the Raven Coloured Progressive Matrices (a multiple-choice IQ test probing abstract reasoning skills). (Sternberg, 'Intelligence, Competence, and Expertise', p. 21.)

42 **As Robert Sternberg watched studies like these pile up – documenting the unusual, sometimes even untestable intelligence traits of Yup'ik Eskimo children, !Kung San hunters of the Kalahari Desert, Brazilian street youth, American horse handicappers, and Californian grocery shoppers – he realised that the lack of correlation between their expertise and IQ scores demanded nothing less than a whole new definition of intelligence.**

—◦≫ Sternberg concludes: 'Abilities as developing forms of expertise [result from] interaction with the demands of the environment.' This was more than seven decades after Sherman and Key had concluded, 'Children develop only as the environment demands development.' (Sternberg, 'Intelligence, Competence, and Expertise', p. 21.)

42 **!Kung San hunters of the Kalahari Desert:** Ceci, *On Intelligence*, p. 35.

42 **Brazilian street youth:** Sternberg, 'Intelligence, Competence, and Expertise', p. 22.

42 **American horse handicappers.**

→ In an utterly fascinating study, Stephen Ceci and his colleague Jeff Liker studied expert and nonexpert horse handicappers at a racetrack. There were two extraordinary findings:

1. 'Even though the greater use of complex, interactive thinking was causally related to success at the racetrack, there was no relation between such complex thinking and IQ or between IQ and success at estimating odds.'

2. Analysis 'was shown to be under the influence of ecological variables such as the sex-role expectations of the task, the physical setting in which the task was performed, the motivational level of the task, and the performance context (game vs. laboratory task).' In other words, environmental variables really mattered. (Ceci, *On Intelligence*, pp. 41–44.)

42 **Californian grocery shoppers:** Sternberg, 'Intelligence, Competence, and Expertise', p. 22.

42 **He saw another problem, too, that reinforced this conclusion: the increasingly flimsy distinction between 'intelligence' tests and so-called achievement tests like the SAT II. The more Sternberg compared the two, the harder it was for him to find any real difference between them.**

Some choice quotes from Sternberg:

> There is no qualitative distinction between various kinds of assessments. The main thing that distinguishes ability tests from achievement tests is not the tests themselves, but rather how psychologists, educators, and others *interpret* the scores on these tests. (Italics mine.)

> Conventional tests of intelligence and related abilities measure achievement that individuals should have accomplished several years back. In other words, the tests are measuring competencies at a somewhat less developed level. Tests such as vocabulary, reading comprehension, verbal analogies, arithmetic problem solving, and the like, are all, in part, tests of achievement. Even abstract reasoning tests measure achievement in dealing with geometric symbols taught in Western schools. One might as well use academic performance to predict ability test scores. The conventional view infers some kind of causation (abilities cause achievement) from correlation, but the inference is not justified from the correlational data.

There is nothing mystical or privileged about the intelligence tests. One could as easily use, say, academic or job performance to predict intelligence-related scores and vice-versa. (Sternberg, 'Intelligence, Competence, and Expertise'.)

43 'Intelligence,' he declared profoundly in 2005, 'represents a set of competencies in development.'

Sternberg calls it 'the model of developing expertise.' (Sternberg, 'Intelligence, Competence, and Expertise', p. 18.)

43 In other words, intelligence isn't fixed. Intelligence isn't general. Intelligence is not a thing. Intelligence is a dynamic, diffuse, and ongoing process.

—⊸⪢ Sternberg argues that no current tests actually measure such built-in intelligence and that intelligence testers are instead relying on a dangerous circular logic: 'Some intelligence theorists point to the stability of the alleged general (*g*) factor of human intelligence as evidence for the existence of some kind of stable and overriding structure of human intelligence. But . . . [w]ith different forms of schooling, *g* could be made either stronger or weaker. In effect, Western forms and related forms of schooling may, in part, create the *g* phenomenon by providing a kind of schooling that teaches in conjunction the various kinds of skills measured by tests of intellectual abilities.'

In other words: we are teaching certain skills in our schools – skills that do correlate reasonably well with Western job performance – and then measuring how well kids learn these skills. Then we pretend that the results reveal a person's raw intelligence, when all they actually reveal is how well a child learned those skills. All we're really learning from intelligence tests is that some kids do better than others in school. We are not, as intelligence testers claim, uncovering the innate cause of these differences.

Is Sternberg saying there's no such thing as innate intelligence?

No. But he is saying that such intelligence is 'not directly measurable', that it is not one general ability which can be scored, and that it is not inherently limiting. The evidence shows that skills and abilities are inextricably interwoven and that all skills are modifiable.

'The main constraint in achieving expertise,' says Sternberg, 'is not some fixed prior level of capacity, but purposeful engagement involving direct instruction, active participation, role modeling, and reward.'

What about the famous correlation between intelligence test scores on the one hand and job performance/life success on the other?

It's a mirage. The correlation does exist, says Sternberg, but not because one causes the other; rather, it's because they both measure the same abilities.

Or as Sternberg puts it: 'Such correlations represent no intrinsic relation between intelligence and other kinds of performance, but rather overlap in the kinds of competencies needed to perform well under different kinds of circumstances. The greater the overlap in skills, in general, the higher the correlations.'

Sternberg then points to a series of studies demonstrating that practical expertise does not correlate well with analytical ('intelligence') tests but *does* correlate very nicely with job performance and life success:

- The Yup'ik Eskimo children of Alaska have 'extremely impressive competencies and even expertise for surviving in a difficult environment, but because these skills are not ones valued by teachers' they tend to do very poorly in school. (*Grigorenko et al.*)
- In Brazil, street children who are extremely successful in running street businesses, and highly expert in maths skills necessary for those affairs, do very poorly in abstract, pencil-and-paper maths problems. (*Nunes*)
- In Berkeley, California, there is 'no correlation' between housewives' impressive abilities in comparison shopping maths and scores on pencil-and-paper maths tests. (*Lave*)

The essential point being that whatever our innate abilities – which clearly exist but are still far from being understood and specified – they do not limit us in a way that IQ scores imply. Ultimately, life success is a function not of inherent abilities, but of highly developed skills.

Sternberg depicts a Western society having painted itself into a logical corner: as we've succeeded with our own brand of academia, we've devised tests – *g*, IQ, SAT, etc. – which we've convinced ourselves show actual innate intelligence, when all they show is achievements according to those particular standards. When you look around the world, you see there are all different kinds of intelligence. Western societies have nothing to be ashamed of in having created successful academies and economies, but we can't let that success corrupt our judgment of where abilities actually come from.

Sternberg: 'Skills develop as results of gene-environment covariation and interaction. If we wish to call them *intelligence*, that is certainly fine, so long as we recognise that what we are calling intelligence is a form of development competencies that can lead to expertise.'

Robert Sternberg, 'Intelligence, Competence, and Expertise'. In *Handbook of Competence and Motivation*, edited by A. J. Elliot and C. S. Dweck, Guilford Publications, 2005.

Grigorenko, Elena. 'The relationship between academic and practical intelligence: a case study of the tacit knowledge of native American Yup'ik people in Alaska'. Office of Educational Research and Improvement, December 2001.

Nunes, T. 'Street Intelligence'. In *Encyclopedia of Human Intelligence*, edited by R. J. Sternberg. Macmillan, 1994, pp. 1045–49.

Lave, J. *Cognition in Practice: Mind, Mathematics, and Culture in Everyday Life*. Cambridge University Press, 1988.

Along the way, a person is not developing a single intelligence, but many different types of intelligence. How many are there? Harvard's Howard Gardner has famously suggested that there are eight different types of intelligence:

Linguistic: the spoken and written word
Logical/mathematical: numbers and reasoning
Musical: rhythm and melody
Spatial intelligence: the ability to form a picture or mental model (highly developed in sailors, engineers, surgeons, sculptors, and painters)
Bodily kinesthetic: intuition and control over one's own body (dancers, athletes, surgeons, craftspeople)
Interpersonal: the ability to understand other people
Intrapersonal: the ability to understand oneself
Naturalist: appreciation and understanding of nature

'Intelligence,' writes Gardner, 'is a biopsychological potential.' It's not an entity, but a living thing. (Gardner, *Intelligence Reframed*, p. 34.)

Or, as Alfred Binet said in 1909: 'With practice, training, and above all method, we manage to increase our attention, our memory, our judgment, and literally to become more intelligent than we were before.' (Binet, *Les idées modernes sur les enfants*, pp. 105–6; this work has been reprinted in Elliot and Dweck, eds., *Handbook of Competence and Motivation*; see p. 124.)

43 'high academic achievers are not necessarily born "smarter"': Csikszentmihályi, Rathunde, and Whalen, *Talented Teenagers*, p. 6.

43 How will that child measure up tomorrow?

→ 'One moves along the continuum,' says Sternberg, 'as one acquires a broader range of skills, a deeper level of the skills one already has, and increased efficiency in the utilisation of these skills.'

Sternberg recalibrated it, in other words, from a thing to a process. The word 'intelligence', he realised, is only a crude symbol for a snapshot of the process in motion. Like any still photograph, it can capture some truth, but it fundamentally misses the ongoing procedure, which is driven, explains Sternberg, by five key elements: metacognitive skills (control of one's own cognition), learning skills, thinking skills, knowledge, and motivation.

Intelligence is not how good you are at something. It's how good you are on your way to becoming.

'At the centre, driving the elements,' observed Sternberg, 'is motivation.' (Sternberg, 'Intelligence, Competence, and Expertise'.)

CHAPTER 3: THE END OF 'GIFTEDNESS' (AND THE TRUE SOURCE OF TALENT)

PRIMARY SOURCES

Eisenberg, Leon. 'Nature, niche, and nurture: the role of social experience in transforming genotype into phenotype'. *Academic Psychiatry* 22 (December 1998): 213–22.

Ericsson, K. Anders. 'Deliberate practice and the modifiability of body and mind: toward a science of the structure and acquisition of expert and elite performance'. *International Journal of Sport Psychology* 38 (2007): 4–34.

Ericsson, K. A., W. G. Chase, and S. Faloon. 'Acquisition of a memory skill'. *Science* 208 (1980): 1181–82.

Howe, Michael J. A., J. W. Davidson, and J. A. Sloboda. 'Innate talents: reality or myth'. *Behavioural and Brain Sciences* 21 (1998): 399–442.

Lehmann, A. C., and K. A. Ericsson. 'The Historical Development of Domains of Expertise: Performance Standards and Innovations in Music'. In *Genius and the Mind*, edited by A. Steptoe. Oxford University Press, 1998, pp. 67–94.

Levitin, Daniel J. *This Is Your Brain on Music: The Science of a Human Obsession*. Dutton, 2006.

CHAPTER NOTES

44 explore the implications of chunking: Chase, *Visual Information Processing*, pp. 215–81.

44 Phone numbers, for example, are not stored in our brains as ten separate numbers but in three easy chunks: 513-673-8754.

—◦⇛ This is my mother's mobile phone number. Call her, tell her how much you like the book so far. Believe me, she won't mind.

45 While our long-term memory capacity is apparently limitless, new memories are almost pathetically fragile: the average healthy adult can reliably juxtapose only three or four new, unrelated items. Such a limit, noted Ericsson and Chase, 'places severe constraints on the human ability to process information and solve problems.'

Seven items are remembered correctly 50 per cent of the time. (Ericsson, Chase, and Faloon, 'Acquisition of a memory skill', pp. 1181–82.)

Excerpt from my earlier book *The Forgetting*, on the importance of a limited memory:

> Why? Why would millions of years of evolution produce a machine so otherwise sophisticated but with an apparent built-in fuzziness, a tendency to regularly forget, repress and distort information and experience?
>
> The answer, it turns out, is that fuzziness is not a severe limitation but a highly advanced feature. As a matter of engineering, the brain does not have any physical limitations in the amount of information it can hold. It is designed specifically to forget most of the details it comes across, so that it may allow us to form general impressions, and from there useful judgements. Forgetting is not a failure at all, but an active metabolic process, a flushing out of data in the pursuit of knowledge and meaning.
>
> We know this not just from brain chemistry and inference, but also because psychologists have stumbled upon a few individuals over the years who actually could not forget enough – and were debilitated by it.
>
> In his *New Yorker* profile, Mark Singer wonders if Martin Scorsese is such a person – burdened by too good a memory. 'Was it, I wondered, painful to remember so much? Scorsese's powers of recall weren't limited to summoning plot turns or notable scenes or acting performances; his grey matter bulged with camera angles, lighting strategies, scores, sound effects, ambient noises, editing rhythms, production credits, data about lenses and film stocks and exposure

speeds and aspect ratios . . . what about all the sludge? An inability to forget the forgettable – wasn't that a burden, or was it just part of the price one paid to make great art?'

For some perspective on the inability to forget, consider the case-study that psychologists call S. In the 1920s, S. was a twenty-something newspaper reporter in Moscow who one day got into trouble with his editor for not taking notes at a staff meeting. In the midst of the reprimand, S. shocked his boss by matter-of-factly repeating everything that had been said in the meeting – word for word.

This was apparently no stretch at all for S., who, it emerged upon closer examination, remembered virtually every detail of sight and sound that he had come into contact with in his entire life. What's more, he took this perfect memory entirely for granted. To him, it seemed perfectly normal that he forgot nothing.

The editor, amazed, sent S. to the distinguished Russian psychologist A. R. Luria for testing. Luria did test him that day, and for many other days over a period of many decades. In all the testing, he could not find any real limit to his capacity to recall details. For example, not only could he perfectly recall tables like this one full of random data after looking at them for just a few minutes –

6	6	8	0
5	4	3	2
1	6	8	4
7	9	3	5
4	2	3	7
3	8	9	1
1	0	0	2
3	4	5	1
2	7	6	8
1	9	2	6
2	9	6	7
5	5	2	0
x	0	1	x

– and not only could he efficiently recite these tables backwards, upside down, diagonal, etc., but after years of memorising thousands of such tables, he could easily reproduce any particular one of them, without warning, whether it was an hour after he had first seen it, or twenty years. The man, it seemed, quite literally remembered everything.

And yet he understood almost nothing. S. was plagued by an inability to make meaning out of what he saw. Unless one pointed the obvious pattern out to him, for example, the following table appeared just as bereft of order and meaning as any other:

1	2	3	4
2	3	4	5
3	4	5	6
4	5	6	7

'If I had been given the letters of the alphabet arranged in a similar order,' he remarked after being questioned about the 1–2–3–4 table, 'I wouldn't have noticed their arrangement.' He was also unable to make sense out of poetry or prose, to understand much about the law, or even to remember people's faces. 'They're so changeable,' he complained to Luria. 'A person's expression depends on his mood and on the circumstances under which you happen to meet him. People's faces are constantly changing; it's the different shades of expression that confuse me and make it so hard to remember faces.'

Luria also noted that S. came across as generally disorganised, dull-witted and without much of a sense of purpose or direction in life. This astounding man, then, was not so much gifted with the ability to remember everything as he was cursed with the inability to forget detail and form more general impressions. He recorded only information, and was bereft of the essential ability to draw meaning out of events. 'Many of us are anxious to find ways to improve our memories,' wrote Luria in a lengthy report on his unusual subject. 'In S.'s case, however, precisely the reverse was true. The big question for him, and the most troublesome, was how he could learn to forget.'

What makes details hazy also enables us to prioritise information, recognise and retain patterns. The brain eliminates trees in order to make sense of, and remember, the forests. Forgetting is a hidden virtue. Forgetting is what makes us so smart. (Shenk, *The Forgetting*, p. 59.)

45 In one-hour sessions, three to five sessions per week, researchers read sequences of random numbers to S.F. at the rate of one digit per second: 2 . . . 5 . . . 3 . . . 5 . . . 4 . . . 9 . . . At intervals, they stopped and asked him to echo their list back. 'If the sequence was reported correctly,' the researchers noted, 'the next sequence was increased by one digit; otherwise it was decreased by one digit.'

Ericsson, Chase, and Faloon write:

> Immediately after half the trials (randomly selected), S.F. provided verbal reports of his thoughts during the trial. At the end of each session, he also recalled as much of the material from the session as he could. On some days, experiments were substituted for the regular sessions. (Ericsson, Chase, and Faloon, 'Acquisition of a memory skill', pp. 1181–82.)

46 From there, the improvements continued unabated: to thirty digits, forty, fifty, sixty, seventy, and finally to a staggering eighty-plus digits before the team concluded the experiment.

—⟫ The 1980 paper says seventy-nine digits in more than 230 hours, but in fact the experiment continued. In the book *Cognitive Skills and Their Acquisition*, they report the higher figures. (Ericsson, Chase, and Faloon, 'Acquisition of a memory skill', pp. 1181–82; Anderson, *Cognitive Skills and Their Acquisition*.)

46 Graph of S.F.'s memory-lab sessions.

> Fig. 1. Average digit span for S.F. as a function of practice. Digit span is defined as the length of the sequence that is correct 50 per cent of the time; under the procedure followed, it is equivalent to average sequence length. Each day represents about 1 hour's practice and ranges from 55 trials per day in the beginning to 3 trials per day for the longest sequences. The 38 blocks of practice shown here represent about 190 hours of practice; interspersed among these practice sessions are approximately 40 hours of experimental sessions (not shown). (Ericsson, Chase, and Faloon, 'Acquisition of a memory skill', pp. 1181–82.)

47 Ericsson and Chase published their results in the prestigious journal *Science*, and their results were subsequently corroborated many times over.

—⟫ In one experimental session, S.F. was switched from digits to letters of the alphabet after three months of practice and exhibited no transfer: his memory span dropped back to about six consonants.

More from that article: 'After all this practice, can we conclude that S.F. increased his short-term memory capacity? There are several reasons to think not.' (Ericsson, Chase, and Faloon, 'Acquisition of a memory skill', pp. 1181–82.)

Google Scholar lists this article as being cited 266 times by other researchers.

47–48 It was a double lesson: when it comes to memory skills, there is no escaping basic human biology – nor any need to. Remembering extraordinary amounts of new information simply requires the right strategies and the right amount of intensive practice, tools theoretically available to any functioning human being.

→ We should acknowledge that evidence from other studies demonstrates that people do arrive at studies with different memory capabilities. 'The conclusion is clear: the talent for being a memory expert reflects both experiential and individual-difference factors. In this case because of the age association and the extreme robustness of the individual difference finding, the likelihood is high that biology based factors are involved.' (Howe, Davidson, and Sloboda, 'Innate talents: reality or myth?', p. 408.)

Relevant studies include:

Anderson, John R. *Cognitive Skills and Their Acquisition.* Lawrence Erlbaum, 1981.

Baltes, Paul B. 'Testing the limits of the ontogenetic sources of talent and excellence'. *Behavioral and Brain Sciences* 21, no. 3 (June 1998): 407–8.

Kliegl, Smith, and P. B. Baltes. 'On the locus and process of magnification of age differences during mnemonic training'. *Developmental Psychology* 26 (1990): 894–904.

It is imperative to understand that I am not arguing against the existence of biological factors or biological differences among individuals. From the moment of conception, everyone has differences. But what has become clear is none of us really know what those biological differences are, or what each of our biological limits really are. When observing our lives in progress, we are not witnessing our biological differences, per se. What we witness even in the early stages of our lives is our life differences resulting from the dynamic interaction of both our unique biologies and our unique environments. The chess game is already in progress, and even after move number three we cannot say that the position on the board was caused by one player's moves.

48 So began Anders Ericsson's remarkable talent odyssey.

→ The stunning results from S.F.'s short-term memory (and a follow-up subject who did even better) led him to suggest a previously unknown memory mechanism called 'long-term working memory' (LT-WM). 'Information in LT-WM is stored in stable form,' he and his coauthor W. Kintsch reported, 'but reliable access to it may be maintained only temporarily by means of retrieval cues in [short-term memory].' They went on to explain:

> In this article we propose that a general account of working memory has to include another mechanism based on skilled use of storage in long-term memory (LTM) that we refer to as long-term working memory (LT-WM) in addition to the temporary storage of information that we refer to as short-term working memory (ST-WM). Information in LT-WM is stored in stable form, but reliable access to it may be maintained only temporarily by means of retrieval cues in ST-WM. Hence LT-WM is distinguished from ST-WM by the durability of the storage it provides and the need for sufficient retrieval cues in attention for access to information in LTM. (Ericsson and Kintsch, 'Long-term working memory', pp. 211–45.)

Ericsson adds:

> Early in the twentieth century it was believed that experts were innately talented with a superior ability to store information in memory. Numerous anecdotes were collected as evidence of an unusual ability to store presented information rapidly. For example, Mozart was supposed to be able to reproduce a presented piece of music after hearing it a single time. However, more recent research has rejected the hypothesis of a generally superior memory in experts and has demonstrated that experts' superior memory is limited to their domains of expertise and can be viewed as the result of acquired skills and knowledge relevant to each specific domain. (Ericsson, 'Superior memory of experts and long-term working memory'.)

48 Though he couldn't be sure at the time, Ericsson suspected he had just discovered the hidden key to the veiled domains of talent and genius. Ericsson writes:

> Experts' superior memory for representative stimuli from their domain of expertise, but not for randomly rearranged versions of those

stimuli, has been frequently replicated in chess (see Charness, 1991, for a review) and also demonstrated in bridge (Charness, 1979; Engle & Bukstel, 1978); go (Reitman, 1976); medicine (G. R. Norman, Brooks & Allen, 1989); music (Sloboda, 1976); electronics (Egan & Schwartz, 1979); computer programming (McKeithen, Reitman, Rueter, & Hirtle, 1981); dance, basketball, and field hockey (Allard & Starkes, 1991); and figure skating (Deakin & Allard, 1991). (Ericsson, 'Superior memory of experts and long-term working memory'.)

48 **Paganini's Sauret cadenza:** From his first violin concerto.

48–49 'Talent' is defined in the *Oxford English Dictionary* as 'mental endowment; natural ability' and is sourced all the way back to the parable of the talents in the book of Matthew.

—⊸⪢ Actually, the word 'talent' goes back much further and was used first for many centuries as a measurement of a weight and then as a name for currency. Its meaning of 'ability' began sometime around its use in the book of Matthew (the parable of the talents, Matthew 25:14–30).

49 The term 'genius', as it is currently defined, goes back to the tail end of the eighteenth century.

Larry Shiner writes:

> At the beginning of the eighteenth century it was widely believed that everyone had a genius or talent for something and that their particular genius could only be perfected by the guidance of reason and rule. By the end of the century, not only had the balance between genius and rule been reversed, but in addition, genius itself had become the opposite of talent and instead of everyone *having* a genius for something, a few people were said to *be* geniuses. (Shiner, *The Invention of Art*, pp. 111–12.)

49 **'Poets and musicians are born,' declared the poet Christian Friedrich Schubart in 1785:** Lowinsky, 'Musical genius', p. 325.

49 **'Musical genius is that inborn, inexplicable gift of Nature,' insisted the composer Peter Lichtenthal in 1826:** Lowinsky, 'Musical genius', p. 324.

49 **'Don't ask, young artist, "what is genius?"' proclaimed Jean-Jacques Rousseau in 1768. 'Either you have it – then you feel it yourself, or you don't – then you will never know it.'**

The longer passage:

> Don't ask, young artist, 'what is genius?' Either you have it – then you feel it yourself, or you don't – then you will never know it. The genius of the musician subjects the entire Universe to his art. He paints all pictures through tones; he lends eloquence even to silence. He renders the ideas through sentiments, sentiments through accents, and the passions he expresses he awakens [also] in his listener's heart. Pleasure, through him, takes on new charms; pain rendered in musical sighs wrests cries [from the listener]. He burns incessantly, but never consumes himself. He expresses with warmth frost and ice. Even when he paints the horrors of Death, he carries in his soul this feeling for Life that never abandons him, and that he communicates to hearts made to feel it. But alas, he does not speak to those who don't carry his seed within themselves and his miracles escape those who cannot imitate them. Do you wish to know whether a spark of this devouring fire animates you? Hasten then, fly to Naples, listen there to the masterworks of Leo, of Durante, of Jommelli, of Pergolesi. If your eyes fill with tears, if you feel your heart beat, if shivers run down your spine, if breath-taking raptures choke you, then take [a libretto by] Metastasio and go to work: his genius will kindle yours; you will create at his example. That is what makes the genius – and the tears of others will soon repay you for the tears that your masters elicited from you. But should the charms of this great artist leave you cold, should you experience neither delirium nor delight, should you find that which transports only 'nice', do you then dare ask what is genius? Vulgar man, don't profane this sublime word. What would it matter to you if you knew it? You would not know how to feel it. Go home and write – French music. (Lowinsky, 'Musical genius', pp. 326–27.)

49 Artists have a vested interest in our believing in the flash of revelation, the so-called inspiration: Lowinsky, 'Musical genius', p. 333.

50 As a vivid illustration, Nietzsche cited Beethoven's sketchbooks.
To see an example of one of Beethoven's working drafts, see Sketches for the 'Pastoral' Symphony (no. 6 in F Major, op. 68). (Ludwig van Beethoven, 1808, British Library Add. MS 31766, f.2.)

50 Beethoven would sometimes run through as many as sixty or seventy different drafts of a phrase before settling on the final one: Wierzbicki, 'The Beethoven Sketch-books'. (Wierzbicki cites Douglas Johnson,

Alan Tyson, and Robert Winter, *The Beethoven Sketchbooks: History, Reconstruction, Inventory*, University of California Press, 1985.)

51 **Over the following three decades, Ericsson and colleagues invigorated the largely dormant field of expertise studies in order to test this idea, examining high achievement from every possible angle: memory, cognition, practice, persistence, muscle response, mentorship, innovation, attitude, response to failure, and on and on. They studied golfers, nurses, typists, gymnasts, violinists, chess players, basketball players, and computer programmers.**

—⊸ A small sampling of their published research, from earliest to most recent:

Conley, D. L., et al. 'Running economy and distance running performance of highly trained athletes'. *Medicine and Science in Sports and Exercise* (1980).

Salthouse, T. A. 'Effects of age and skill in typing'. *Journal of Experimental Psychology: General* (1984).

Schulz, R., et al. 'Peak performance and age among superathletes: track and field, swimming, baseball, tennis, and golf'. *Journal of Gerontology* (1988).

Coyle, E. F., et al. 'Physiological and biomechanical factors associated with elite endurance cycling performance'. *Medicine and Science in Sports and Exercise* (1991).

Abernethy, B., et al. 'Visual-perceptual and cognitive differences between expert, intermediate, and novice snooker players'. *Applied Cognitive Psychology* (1994).

Starkes, J. L., et al. 'A new technology and field test of advance cue usage in volleyball'. *Research Quarterly for Exercise and Sport* (1995).

Krampe, R. Th., et al. 'Maintaining excellence: deliberate practice and elite performance in young and older pianists'. *Journal of Experimental Psychology* (1996).

Higbee, K. L. 'Novices, apprentices, and mnemonists: acquiring expertise with the phonetic mnemonic'. *Applied Cognitive Psychology* (1997).

Nevett, M. E., et al. 'The development of sport-specific planning, rehearsal, and updating of plans during defensive youth baseball game performance'. *Research Quarterly for Exercise and Sport* (1997).

Masters, K., et al. 'Associative and dissociative cognitive strategies in exercise and running: 20 years later, what do we know?' *Sport Psychologist* (1998).

Pieper, H.-G. 'Humeral torsion in the throwing arm of handball players'. *American Journal of Sports Medicine* (1998).

Gabrielsson, A. 'The Performance of Music'. In *The Psychology of Music*, edited by D. Deutsch. Academic Press, 1999.

Helson, W. F., et al. 'A multidimensional approach to skilled perception and performance in sport'. *Applied Cognitive Psychology* (1999).

Helgerud, J., et al. 'Aerobic endurance training improves soccer performance'. *Medicine and Science in Sports and Exercise* (2001).

Hopkins, W. G., et al. 'Variability of competitive performance of distance runners'. *Medicine and Science in Sports and Exercise* (2001).

Pelliccia, A., et al. 'Remodeling of left ventricular hypertrophy in elite athletes after long-term deconditioning'. *Circulation* (2002).

Goldspink, G. 'Gene expression in muscle in response to exercise'. *Journal of Muscle Research and Cell Motility* (2003).

Maguire, E. A., et al. 'Routes to remembering: the brains behind superior memory'. *Nature Neuroscience* (2003).

McPherson, S., et al. 'Tactics, the neglected attribute of expertise: problem representations and performance skills in tennis'. In *Expert Performance in Sports*, edited by Janet Starkes and K. Anders Ericcson. Human Kinetics Publishers, 2003.

Pantev, C., et al. 'Music and learning-induced cortical plasticity'. *Annals of the New York Academy of Sciences* (2003).

Duffy, L. J., B. Baluch, and K. A. Ericsson. 'Dart performance as a function of facets of practice amongst professional and amateur men and women players'. *International Journal of Sports Psychology* 35 (2004): 232–45.

Ericsson, K. A. 'Deliberate practice and the acquisition and maintenance of expert performance in medicine and related domains'. *Academic Medicine* (2004).

Prior, B. M., et al. 'What makes vessels grow with exercise training?' *Journal of Applied Physiology* (2004).

Pyne, D. B., et al. 'Progression and variability of competitive performance of Olympic swimmers'. *Journal of Sports Sciences* (2004).

Wittwer, M., et al. 'Regulatory gene expression in skeletal muscle of highly endurance trained humans'. *Acta Physiologica Scandinavica* (2004).

Baker, J., et al. 'Cognitive characteristics of expert, middle of the pack, and back of the pack ultra-endurance triathletes'. *Psychology of Sport and*

Exercise (2005).

Bengtsson, S. L., et al. 'Extensive piano practicing has regionally specific effects on white matter development'. *Nature Neuroscience* (2005).

Larsen, H., et al. 'Training response of adolescent Kenyan town and village boys to endurance running'. *Scandinavian Journal of Medicine and Science in Sports* (2005).

Legaz, A., et al. 'Changes in performance, skinfold thicknesses, and fat patterning after three years of intense athletic conditioning in high level runners'. *British Journal of Sports Medicine* (2005).

van der Maas, H. L. J., et al. 'A psychometric analysis of chess expertise'. *American Journal of Psychology* (2005).

Young, L., et al. 'Left ventricular size and systolic function in Thoroughbred race horses and their relationship to race performance'. *Journal of Applied Physiology* (2005).

Coffey, V. G., et al. 'Interaction of contractile activity and training history on mRNA abundance in skeletal muscle from trained athletes'. *American Journal of Physiology, Endocrinology, and Metabolism* (2006).

51 **His father, Leopold Mozart, was an intensely ambitious Austrian musician.**

Leopold's book was published the year his son Wolfgang was born. (Sadie, ed., *The Grove Concise Dictionary of Music*, 1988.)

52 **[Leopold] advocated the so-called 'Geminiani grip':** November, 'A French edition of Leopold Mozart's *Violinschule* (1756)'.

52 **Then came Wolfgang. Four and a half years younger than his sister, the tiny boy got everything Nannerl got – only much earlier and even more intensively.**

—◦⪢ There is a wonderful parallel with another family three centuries later – the three Polgar sisters in Hungary, all raised to be exceptional chess players. As each girl was exposed to chess earlier than her elder sister, she subsequently became the better player. The youngest, Judit, became the youngest grandmaster in history at age fifteen (at that time). (Shenk, *The Immortal Game*, p. 132.)

52–53 **Literally from his infancy, he was the classic younger sibling soaking up his big sister's singular passion. As soon as he was able, he sat beside her at the harpsichord and mimicked chords that she played.**

—◦⪢ Nannerl later wrote: 'He often spent much time at the clavier [keyboard], picking out thirds . . . and his pleasure showed it sounded good [to him].' (Zaslaw and Cowdery, *The Compleat Mozart*, p. 276.)

His sister also echoed her father's words that Wolfgang was the beneficiary of a God-given talent, and that his abilities were apparent from very early on. This may seem to be in contradiction with my argument here. But neither the intense religiosity of the Mozart family nor the obvious precociousness of young Wolfgang refutes the notion of his genius being a matter of development.

53 **Not only did Leopold openly give preferred attention to Wolfgang over his daughter; he also made a career-altering decision to more or less shrug off his official duties in order to build an even more promising career for his son.**

'Everything connected with his son's career was of such importance that his official duties fell into the background.' (Geiringer, 'Leopold Mozart', pp. 401–4.)

Also, Alfred Einstein writes:

> Up to 1762, [Leopold's] ambition to rise in Salzburg to the highest position had been thwarted by his superior, the Kapellmeister Johann Ernst Eberlin, who towered far above him as a creative musician, and whom he himself recognised as a pattern 'of a thorough and finished master,' as an example of wonderful fertility and ease of production. But some months before Eberlin died (1762), Leopold had departed with his children on his second tour which, as a moral obligation and as a pecuniary speculation, he put far above his official duties at Salzburg. (Alfred Einstein, preface to *A Treatise on the Fundamental Principles of Violin Playing*, p. xvii. See also Stowell, 'Leopold Mozart Revised', pp. 126–57, and November, 'A French Edition of Leopold Mozart's *Violinschule* [1756]'.)

53 **From the age of three, then, Wolfgang had an entire family driving him to excel with a powerful blend of instruction, encouragement, and constant practice.**

—◦➾ Have we identified *every* explanation for the marvelous success of Leopold Mozart's children? Of course not. This book does not pretend that there is a simple recipe for talent or presume to fully understand the dynamic that makes the children of some ambitious parents into amazingly skilled performers and others mediocre or disinterested players. The point here is that it is a dynamic process – not that we can track every single factor and interaction as it plays out.

53 **today many young children exposed to Suzuki and other rigorous musical programmes play as well as young Mozart did – and some play even better:** Lehmann and Ericsson, 'The Historical Development of Domains of Expertise', pp. 67–94.

—∘⋟ Deconstructing the myth of Mozart's early achievements, and understanding why they were so rare, does not make those achievements any less spectacular. It is a blessing for anyone, at any age, to be able to bring grace and beauty into other people's lives. For a child to attain such poise and proficiency while his or her peers lark about on swings and fumble with toy instruments is truly something to behold.

That having been said, no one today would pay any attention to Mozart's earliest years if he hadn't gone on to develop into such a remarkable adult composer.

54 **'When we say that someone is talented':** Levitin, *This Is Your Brain on Music*, p. 196.

55 **Practice changes your body. Researchers have recorded a constellation of physical changes (occurring in direct response to practice) in the muscles, nerves, hearts, lungs, and brains of those showing profound increases in skill level in any domain.**

Ericsson writes:

> [There is] emerging evidence that extended focused practice has profound effects on, and can influence virtually every aspect of, the human body, such as muscles, nerve systems, heart and circulatory system, and the brain. (Ericsson et al., eds., *The Cambridge Handbook of Expertise and Expert Performance*, p. 59.)

55 **The brain drives the brawn. Even among athletes, changes in the brain are arguably the most profound, with a vast increase in precise task knowledge, a shift from conscious analysis to intuitive thinking (saving time and energy), and elaborate self-monitoring mechanisms that allow for constant adjustments in real time.**

—∘⋟ Supporting Ericsson's thesis is his observation, from many pieces of research, that 'expert performance is *primarily* mediated by acquired mental representations that allow the experts to anticipate courses of action, to control those aspects that are relevant to generating their superior performance, and to evaluate alternative courses of action during performance or after the completion of the competition.' (Italics mine.) (Ericsson, 'Deliberate practice and the modifiability of body and mind', pp. 4–34.)

In other words, most of the advantages held by superior achievers, even among athletes, occur in particular regions of the brain. Better musicians, typists, hockey goalies, etc., are all able to draw on more elaborate mental representations of what they want to do – and to execute them more efficiently.

This first came to researchers' attention in studies of typists, when the researchers realised that the better and faster typists were able to look further ahead and prepare themselves better for future keystrokes. Later, they observed the same thing with hockey goalies, tennis players, and baseball batters – showing that they had more elaborate mental preparation for the events about to unfold *and* that they could more efficiently draw on better 'anticipatory cues' to make better decisions and execute more efficient motor function in real time.

'Experts certainly know more, but they also know differently,' says Ericsson. 'Expertise is . . . not a simple matter of fact or skill acquisition, but rather a complex construct of adaptations of mind and body, which include substantial self-monitoring and control mechanisms.'

He continues: 'There is an element of unencumbered elegance in expert performance, the underpinnings of which are based on the efficient management and control of the adaptive processes. A source for this might be in abstracted layers of control and planning.' (Ericsson et al., eds., *The Cambridge Handbook of Expertise and Expert Performance*, p. 57.)

56 **'Deliberate practice is a very special form of activity':** Ericsson et al., 'Giftedness and evidence for reproducibly superior performance', pp. 3–56.

56 **Recall Eleanor Maguire's 1999 brain scans of London cabbies, which revealed greatly enlarged representation in the brain region that controls spatial awareness. The same holds for any specific task being honed; the relevant brain regions adapt accordingly.**

See earlier note on page 166, beginning, 'Further, her conclusion was perfectly consistent with what others have discovered in recent studies . . .'

57 **Whereas the amateur singers experienced the lesson as self-actualisation and an enjoyable release of tension, the professional singers increased their concentration and focused on improving their performance during the lesson:** Ericsson, K. Anders, Roy W. Roring, and Kiruthiga Nandagopal. 'Giftedness and evidence for reproducibly superior performance: an account based on the expert performance framework'. *High Ability Studies* 18, no. 1 (June 2007): 3–56.

The same phenomenon is discussed in the following works:

Charness, Neil, R. Th. Krampe, and U. Mayr. 'The Role of Practice and Coaching in Entrepreneurial Skill Domains: An International Comparison of Life-Span Chess Skill Acquisition'. In *The Road to Excellence: The Acquisition of Expert Performance in the Arts and Sciences, Sports, and Games*, edited by K. A. Ericsson. Lawrence Erlbaum, 1996, pp. 51–80.

Charness, Neil, M. Tuffiash, R. Krampe, E. Reingold, and E. Vasyukova. 'The role of deliberate practice in chess expertise'. *Applied Cognitive Psychology* 19 (2005): 151–65.

Duffy, L. J., B. Baluch, and K. A. Ericsson. 'Dart performance as a function of facets of practice amongst professional and amateur men and women players'. *International Journal of Sports Psychology* 35 (2004): 232–45.

Ward, P., N. J. Hodges, A. M. Williams, and J. L. Starkes. 'Deliberate Practice and Expert Performance: Defining the Path to Excellence'. In *Skill Acquisition in Sport: Research, Theory and Practice*, edited by A. M. Williams and N. J. Hodges. Routledge, 2004.

58 **Genes are involved, of course. They're a dynamic part of the process as they become activated.**

Ericsson writes:

> The adult body has evolved to cope with short-term fluctuations in physiological demands . . . Whenever individuals engage in physical sport activities, the metabolism of their muscle fibres increases, and the supply of oxygen and energy within the muscle cells is rapidly reduced and supplies are extracted from the nearest blood vessels. To preserve homeostasis, the body activates various counter measures (negative feedback loops). For example, increased breathing rates increase oxygen concentrations and decrease carbon dioxide concentrations in the blood. In turn, the conversion of stored energy replenishes expendable energy available in the blood, and the increased rate of blood circulation distributes these commodities to the systems of the body with the greatest needs. However, when individuals deliberately push themselves beyond the zone of relative comfort (Ericsson, 2001, 2002) and engage in sustained strenuous physical activity, they will challenge the available protection of homeostasis sufficiently to induce an abnormal state for cells in some physiological systems. These states will sometimes be associated with abnormally low levels of certain vital elements and compounds, such as

> oxygen, and energy-related compounds (e.g., glucose, adenosine-diphosphate; ADP and adenosine-triphosphate; ATP), which lead metabolic processes to change and produce alternative biochemical products. These biochemical states will trigger the activation of some genes in massive storage of dormant genes within the cells' DNA. The activated genes in turn will stimulate and 'turn on' biochemical systems designed to cause bodily reorganisation and adaptive change. Recent research shows that the biochemical response of cells to various types of strain induced by vigorous activity, such as physical exercise, is very complex. Even more directly relevant to physical exercise, over one hundred different genes are activated and expressed in mammalian muscle in response to intense physical exercise. (Ericsson, 'Giftedness and evidence for reproducibly superior performance', pp. 3–56.)

58 **'When individuals deliberately push themselves':** Ericsson, 'Giftedness and evidence for reproducibly superior performance', pp. 3–56.

58 **This does not mean, of course, that every person has the same resources and opportunity, or that anyone can be great at anything; biological and circumstantial differences and advantages/disadvantages abound. But in revealing talent to be a process, the simple idea of genetic giftedness is forever debunked. It is no longer reasonable to attribute talent or success to a specific gene or any other mysterious gift.**

Ericsson writes:

> A careful review of the published evidence on the heritability of acquisition of elite sports achievement failed to reveal reproducible evidence for any genetic constraints for attaining elite levels by healthy individuals (excluding, of course, the evidence on body size). (Ericsson, 'Deliberate practice and the modifiability of body and mind', pp. 4–34.)

R. Subotnik adds:

> In order to be gifted, that is, to be exceptional, as one matures, one needs to be increasingly active in one's own development. You have to develop your hunger, you have to be open to career advice, and you have to hone your social skills or your intriguing persona. (Subotnik, 'A developmental view of giftedness', pp. 14–15.)

59 **From sublime pianists to unusually profound physicists, researchers have been very hard-pressed to find any examples of truly**

extraordinary performers in any field who reached the top of their game before that ten-thousand-hour mark.

Daniel Levitin writes:

> In study after study, of composers, basketball players, fiction writers, ice-skaters, concert pianists, chess players, master criminals . . . this number comes up again and again. Ten thousand hours is equivalent to roughly three hours a day, or 20 hours a week, of practice over 10 years . . . No one has yet found a case in which true world-class expertise was accomplished in less time. It seems that it takes the brain this long to assimilate all that it needs to know to achieve true mastery. (Levitin, *This Is Your Brain on Music*, p. 193.)

Recent chess studies conform with Levitin's and Ericsson's observations in a number of ways – practice hours, starting age, etc. (Campitelli and Gobet, 'The role of practice in chess'; Gobet and Campitelli, 'The role of domain-specific prac-tice, handedness and starting age in chess', pp. 159–72.)

59 'People make a great mistake who think that my art has come easily to me,' Mozart himself once wrote to his father, as if to make this precise point. 'Nobody has devoted so much time and thought to compositions as I.'

He continues: 'There is not a famous master whose music I have not studied over and over.' (Pott, 'The Triumph of Genius'.)

59 His first seven piano concertos, written from ages eleven to sixteen, 'contain almost nothing original,' reports Temple University's Robert Weisberg, and 'perhaps should not even be labelled as being by Mozart.'

—⟶ And they may not even have been that impressive – they exist today only in his father's handwriting.

Robert W. Weisberg writes:

> Mozart seems to have begun learning his skill through study and small-scale modification of the works of others. Mozart arranged it for piano and other instruments . . . Even when Mozart began to write music of his own, those pieces seem to have been based relatively closely on works by other composers, as can be seen in his production of symphonies. (Weisberg, 'Case Studies of Innovation', p. 214.)

Jon Pott adds: 'Many of his early compositions were dazzling and accomplished for his age, but not for more.' Pott also writes that critics consider his Symphony no. 29, written ten years after his first symphony, to be his first work of real stature. (Pott, 'The Triumph of Genius'. See also Weisberg, 'Expertise in Creative Thinking', pp. 761–87.)

CHAPTER 4: THE SIMILARITIES AND DISSIMILARITIES OF TWINS

PRIMARY SOURCES

Bateson, Patrick. 'Behavioral Development and Darwinian Evolution'. In *Cycles of Contingency: Developmental Systems and Evolution*, edited by Susan Oyama et al. MIT Press, 2003.

Bateson, Patrick, and Paul Martin. *Design for a Life: How Biology and Psychology Shape Human Behavior*. Simon & Schuster, 2001.

Downes, Stephen M. 'Heredity and Heritability'. Published online on the Stanford Encyclopedia of Philosophy Web site, first posted July 15, 2004; revised May 28, 2009.

Joseph, Jay. *The Gene Illusion: Genetic Research in Psychiatry and Psychology under the Microscope*. Algora Publishing, 2004.

Moore, David S. *The Dependent Gene: The Fallacy of 'Nature vs. Nurture'*. Henry Holt, 2003.

Ridley, Matt. *Nature via Nurture*. HarperCollins, 2003.

Turkheimer, Eric, Andreana Haley, Mary Waldron, Brian D'Onofrio, and Irving I. Gottesman. 'Socioeconomic status modifies heritability of IQ in young children'. *Psychological Science* 14, no. 6 (November 2003): 623–28.

CHAPTER NOTES

61 Ted Williams retired from baseball on September 28, 1960, at age forty-two.

Standing before a grateful hometown crowd at Fenway Park and facing Baltimore's Jack Fisher on the mound. (Full game stats available online at baseball-reference.com.)

62 'What if we could sell dad's DNA and there could be little Ted Williamses all over the world?': Farrey, 'Awaiting Another Chip off Ted Williams' Old DNA?'

62 Rainbow the cat and her clone Cc.

Kristen Hays writes:

> Rainbow the cat is a typical calico with splotches of brown, tan and gold on white. Cc, her clone, has a striped grey coat over white. Rainbow is reserved. Cc is curious and playful. Rainbow is chunky. Cc is sleek . . . Sure, you can clone your favourite cat. But the copy will not necessarily act or even look like the original. (Hays, 'A Year Later, Cloned Cat Is No Copycat: Cc Illustrates the Complexities of Pet Cloning'.)

63 **'Identical genes don't produce identical people':** Wray, Sheler, and Watson, 'The World After Cloning', pp. 59–63.

63 **'In theory, you could create someone who would be a step ahead of other people':** Farrey, 'Awaiting Another Chip off Ted Williams' Old DNA?'

64 **Coincidentally, they'd been given the same first name by their adoptive parents.**

—⊸ In fact, they had the same first name and virtually the same middle name: James Alan Lewis and James Allen Springer. These were names given separately by adopted parents, which could only reflect culture or coincidence, not genetics – but it does play to the eerie magical quality of the story.

64 **'I thought we were going to do a single case study,' Bouchard later recalled:** Wright, *Twins*, p. 46.

64 **'Nothing seems to me more curious,' he once wrote, 'than the similarity and dissimilarity of twins':** Charles Darwin, in a letter to Francis Galton, November 7, 1875, as published on the Galton.org Web site.

64 **Since identical twins were thought to share 100 per cent of their DNA.**

—⊸ In fact, identical twins turn out not to have exactly the same DNA. Very close, but not exactly the same. (Anahad O'Connor, 'The Claim: Identical Twins Have Identical DNA', *New York Times*, March 11, 2008.)

65 **Journalists were understandably blown away when Bouchard and colleagues published data that seemed to demonstrate that genes were responsible for roughly 60 per cent of intelligence, 60 per cent of personality, 40–66 per cent of motor skills, 21 per cent of creativity.**

INTELLIGENCE

Herrnstein, Richard J., and Charles Murray. *The Bell Curve*. Free Press, 1994, p. 298. The authors average a number of estimates from 40–80 %.

PERSONALITY

Bouchard, T. J., Jr., and Yoon-Mi Hur. 'Genetic and environmental influences on the continuous scales of the Myers-Briggs type indicator: an analysis based on twins reared apart'. *Journal of Personality* 66, no. 2 (2008): 135.

MOTOR SKILLS

Fox, Paul W., Scott L. Hershberger, and Thomas J. Bouchard Jr. 'Genetic and environmental contributions to the acquisition of a motor skill'. *Nature* 384 (1996): 356.

CREATIVITY

Nichols, R. 'Twin studies of ability, personality, and interests'. *Homo* 29 (1978): 158–73.

65 **'Since personality is heritable . . .' (*New York Times*):** Nicholas Wade, 'The Twists and Turns of History, and of DNA', *New York Times*, March 12, 2006.

65 **'Men's Fidelity Controlled by "Cheating Genetics"' (Drudge Report):** Drudge Report, September 3, 2008.

Also: 'Forty per cent of [marital] infidelity [can] be blamed on genes.' (Highfield, 'Unfaithful?')

65–66 **'The genetic idea has had a tumultuous passage through the twentieth century,' he wrote, 'but the prevailing view of human nature at the end of the century resembles in many ways the view we had at the beginning . . . Circumstances do not so much dictate the outcome of a person's life as they reflect the inner nature of the person living it. Twins have been used to prove a point, and the point is that we don't become. We are':** Wright, *Twins*, p. 10.

—⊸➾ This is really an extraordinary and very unfortunate statement. Lawrence Wright is a distinguished journalist and writer, and I am an admirer. But even great journalists and scientists can get caught up in misinterpreted science, and that's what appears to have happened in this case.

66 **Turkheimer found that intelligence was not 60 per cent heritable, nor 40 per cent, nor 20 per cent, but *near* 0 per cent.**

'The models suggest,' Turkheimer wrote, 'that in impoverished families, 60% of the variance in IQ is accounted for by the shared environment, and the contributions of genes *is close to zero*; in affluent families, the result is almost exactly the reverse.' (Italics mine.) (Turkheimer et al., 'Socioeconomic status modifies heritability of IQ in young children', p. 632.)

66 **'a model of [genes plus environment] is too simple':** Turkheimer et al., 'Socioeconomic status modifies heritability of IQ in young children', p. 627.

67 **Heritability, explains author Matt Ridley, 'is a population average, meaningless for any individual person':** Ridley, *Nature via Nurture*, p. 76.

68 **Early shared GxE. Identical twins share a wide collection of similarities not just because they share the same genes, but because they share the same genes *and* early environments – hence, the same gene-environment interactions throughout gestation.**

—∘⪢ In addition to nine months of shared prenatal environment, most also have some weeks or months of shared postnatal environment before separation.

68 **Shared cultural circumstances. In identical twins comparisons, shared biology always grabs all the attention. Inevitably overlooked is the vast number of shared cultural traits: same age, same sex, same ethnicity, and, in most cases, a raft of other shared (or very similar) social, economic, and cultural experiences.**

> The mere fact that two people are born on the same day can have an important impact on their subsequent behaviour and beliefs. (Joseph, *The Gene Illusion*, p. 105.)

68 **'All of these factors work towards increasing the resemblance of reared-apart twins,' explains psychologist Jay Joseph.**

For other psychologists not to recognise their importance, he argues, is a 'stunning failure'. (Joseph, *The Gene Illusion*, p. 100.)

68 **To test the influence of just a few of them, psychologist W. J. Wyatt assembled fifty college students completely unrelated and unknown to**

one another and then placed them in random pairings purely on the basis of age and sex: Joseph, *The Gene Illusion*, p. 100; Wyatt, Posey, Welker, and Seamonds, 'Natural levels of similarities between identical twins and between unrelated people', p. 64.

68–69 Hidden dissimilarities. Statisticians call it 'the multiple-end-point problem': the seductive trap of selectively picking data that fit a certain thesis, while conveniently discarding the rest. For every tiny similarity between the Jim twins, there were thousands of tiny (but unmentioned) dissimilarities. 'There are endless possibilities for doing bad statistical inferences,' says Stanford statistician Persi Diaconis. 'You get to pick which features you want to resonate to. When you look at your mom, you might say, "I'm exactly the opposite." Someone else might say, "Hmm."'

Gina Kolata adds: 'And when we look at our parents, or our children, and see ourselves, it is easier than we think to get caught in the multiple-end-points statistical trap.' (Kolata, 'Identity'.)

69 *New York Times* science writer Natalie Angier adds: 'What the public doesn't hear of are the many discrepancies between the twins. I know of two cases in which television producers tried to do documentaries about identical twins reared apart but then found the twins so distinctive in personal style – one talky and outgoing, the other shy and insecure – that the shows collapsed of their own unpersuasiveness': Angier, 'Separated by Birth?'

These separated-twin stories, added behaviour geneticist Richard Rose, '[make] good show biz but uncertain science.' (Joseph, *The Gene Illusion*, p. 107.)

Jay Joseph adds:

> Judith Harris has written that 'there are too many of these stories for them all to be coincidences,' and it is true – they are not coincidences; they are selectively reported 'show biz' combined with a stunning failure to recognise the environmental factors influencing these twins' similar behaviours. (Joseph, *The Gene Illusion*, p. 107.)

69 Coordination and exaggeration. All twins feel a close bond with each other, and while child twins growing up together might often cling to their differences, reunited adult twins understandably revel

in their similarities. Researchers try to guard against any purposeful or unwitting coordination, but in her 1981 book *Identical Twins Reared Apart*, Susan Farber reviewed 121 cases of twins described by researchers as 'separated at birth' or 'reared apart'. Only three of those pairs had actually been separated shortly after birth *and* studied at their first reunion.

—⊸⪢ Were these studied twins truly separate? Susan Farber reviewed 121 cases in her 1981 book *Identical Twins Reared Apart* – only three pairs had been truly separated shortly after birth and studied at their first reunion.

Consider also the case of Oskar Stöhr and Jack Yufe, perhaps the most compelling reunited twins ever. The identical twins were separated shortly after birth by their divorced parents, the former raised in Nazi Germany, the latter raised as a Jew in Trinidad. Despite the obvious cultural differences, their reunion at age forty-seven stunned the world with similarities: wire-rimmed glasses, moustaches, two-pocket shirts, love of spicy foods and sweet liquors, absentmindedness, habits of sleeping in front of the TV and flushing the toilet before using it. Their reported similarities were astounding indeed – until one realised that they had already been in contact for twenty-five years.

Another entertaining twosome earned the nickname 'Giggle Sisters' for their constant and similar laugh. They were also both frugal, shared blue as a favourite colour, drank their coffee black and cold, 'squidged' up their noses, had once worked as polling clerks, and had each suffered a miscarriage with their first pregnancy. After being interviewed by researchers, though, the Giggle Sisters acknowledged inventing at least one shared life goal. (Joseph, *The Gene Illusion*, p. 100; Farber, *Identical Twins Reared Apart*, p. 100.)

Bouchard reported that the average age of his twins studied was forty, with an average of thirty years spent apart – meaning that there was an average of ten years of contact. (Wright, *Twins*, p. 69.)

69 Considering all this, was it really so shocking that Jim Lewis and Jim Springer, two thirty-nine-year-old men who shared a womb for nine months and a month more in the same hospital room, *and* were raised in working-class towns seventy miles apart (by parents with tastes similar enough to name their kids Jim and Larry), would end up preferring the same beer, same cigarettes, same car, same hobbies, and have some of the same habits?

—⊸⪢ Do you, reader, perhaps have a 'cultural twin' out there who you've never met? Someone the same age from your same hometown who shares a few of your food passions, music passions, etc.? I grew up in Cincinnati, Ohio, in the 1970s. I wonder how hard it would be to find a forty-two-year-old I've never met from the same region who today likes Bruce Springsteen, Graeter's ice cream, and Porsche cars, who plays the acoustic guitar, and who lost interest in baseball after Pete Rose left the Cincinnati Reds. I'd wager I could find one on the streets of Cincinnati in about three minutes. We could probably fill a baseball stadium with us . . .

69 Lest anyone think they were living perfectly parallel lives: Chen, 'Twins Reared Apart'.

70 Otto (left) and Ewald (right).
Michael Rennie writes:

> Since the sequencing of the human genome there has been an expectation that we will be able to unveil many of the secrets underlying ways in which the human body is put together, the differences that exist between individuals in muscle and bone mass and composition, and how adaptable they are to physical activity. Although there have been some successes in identifying genes that are associated with particular musculoskeletal functions, it seems that, as for many other human attributes, human body size and composition are as much a matter of environment as of natural endowment, with each having about 50% influence. The gentlemen pictured in Fig. 1 [Otto and Ewald] are in fact identical twins who chose to sculpt their bodies by different training regimes to completely different results, in order to pursue athletic careers in distance running and field events. Obviously the scope for environmental effects is large. Most of what I will discuss concerns relatively short-term effects of food and exercise, i.e. those which occur within a time frame of up to 72 h, and I am going to say very little about alterations of gene transcription, since this has not been the focus of our work until recently. Nevertheless, it did come as a surprise to me and other workers to realise that it was possible to see marked alterations in gene expression within 2 h of finishing a bout of exercise or infusing insulin; given the much slower metabolic rate of human organs compared to that of a rat or a mouse, it was to be expected that these changes would take much longer. (Rennie, 'The 2004 G. L. Brown Prize Lecture', pp. 427–28.)

Art De Vany writes:

It turns out that Otto's more low intensity stimulation decreased ATP concentrations and activated AMP kinase. This inhibited stimulation of TSC2 so that mTOR-mediated myofibrillar stimulation did not occur. In Ewald's case, the genes got another signal: high intensity contraction stimulated PKB activity, increasing TSC2 and activating the mTOR signal, resulting in markedly increased myofibrillar protein synthesis.

So, a low intensity signal turns on different genes and signal cascades than a high intensity signal. Low intensity – no muscle protein synthesis. High intensity – markedly increased muscle protein synthesis. Same genes, different signals, different bodies. (De Vany, 'Twins'.)

CHAPTER 5:
PRODIGIES AND LATE BLOOMERS

PRIMARY SOURCES

Halberstam, David. *Playing for Keeps*. Broadway Books, 2000.

Hulbert, Ann. 'The Prodigy Puzzle'. *New York Times*, November 20, 2005.

Levitin, Daniel J. *This Is Your Brain on Music: The Science of a Human Obsession.* Dutton, 2006.

Ma, Marina. *My Son, Yo-Yo*. Chinese University Press, 1996.

Terman, Lewis M. 'The Discovery and Encouragement of Exceptional Talent'. Walter Van Dyke Bingham Lecture at the University of California, Berkeley, March 25, 1954.

Terman, Lewis M. *Genetic Studies of Genius*. Stanford University Press.

Volume I: *Mental and Physical Traits of a Thousand Gifted Children* (1925).

Volume II: *The Early Mental Traits of Three Hundred Geniuses* (1926).

Volume III: *The Promise of Youth, Follow-up Studies of a Thousand Gifted Children* (1930).

Volume IV: *The Gifted Child Grows Up* (1947).

Volume V: *The Gifted Group at Mid-Life* (1959).

Winner, Ellen. 'The origins and ends of giftedness'. *American Psychologist* 55, no. 1 (2000): 159–69.

CHAPTER NOTES

71 **They called it 'hang time'.**

The fascination with Jordan's flight became so deep that after a while, physicists felt compelled to jump in and reassure people that Jordan was not, in fact, defying gravity.

'By bringing his knees up, he's raising his centre of mass relative to his head,' explained Michael Kruger, chairman of physics at the University of Missouri–Kansas City. 'He does that on his way up. On the way down, of course, he lowers his legs and that lowers his centre of mass which is bringing it back to where it normally is, which effectively raises his head relative to the centre of mass. The head no longer follows the parabola. The head stays up there at one height. So what you get is during the entire time, the head stays at the same height. The centre of mass goes up and down, through gravity and him manipulating his centre of mass.

'When we look at each other, we don't intuitively know where our centre of mass is. We fixate on things, like the head. But this really is happening; the head is staying constant for an unnaturally long time because he manipulates his centre of mass.' (Grathoff, 'Science of Hang Time'.)

The American Association of Physics Teachers provides this explanation:

> How high someone can jump depends on the force used to push on the floor when starting to jump, which in turn depends on the strength and power of the jumper's leg muscles. The harder and more powerful the jump, the higher and longer the flight. In order to leap four feet into the air, the hang time would be 1.0 seconds. Jordan had a few tricks up his sleeve to make that hang time seem longer. When he dunked, he held onto the ball a bit longer than most players, and actually placed it in the basket on the way down. He also pulled his legs up as the jump progressed so it appeared that he was jumping higher. But it still all happened in less than one second. (American Association of Physics Teachers, 'Slam Dunk Science'.)

71 **'pure genius is something very, very rare':** Halberstam, *Playing for Keeps*, p. 9.

72 **'If Michael Jordan was some kind of genius, there had been few signs of it when he was young':** Halberstam, *Playing for Keeps*, p. 17.

72 **Yo-Yo Ma, on the other hand, showed his stuff from very early on:** Ma, *My Son, Yo-Yo*.

72 **Pablo Casals called him simply 'Wonder Boy':** Ma, *My Son, Yo-Yo*, p. 80.

72 **researchers have discovered that child prodigies and adult superachievers are very often not the same people. For every wonder child Yo-Yo Ma who also thrives in adulthood, there is a long list of child prodigies who never become remarkable adult achievers.**

—•» 'Most gifted children, even most child prodigies, do not go on to become adult creators,' says Boston College's Ellen Winner. (Winner, 'The origins and ends of giftedness', pp. 159–69.)

72 **At the same time, an equally long list of profound adult achievers manage to attain greatness without first showing any profound abilities as children – a list that includes Copernicus, Rembrandt, Bach, Newton, Kant, da Vinci, and Einstein.**

—•» This list comes from Malcolm Gladwell, in a talk he gave to the Association for Psychological Science in 2006. (Wargo, 'The myth of prodigy and why it matters'.)

San Jose State University psychologist Gregory Feist adds: 'Early childhood talent in music by no means is a necessary or a sufficient condition for adult creative achievement. It is often the case that the musically most-accomplished adults do not begin to set themselves apart in any significant way until middle adolescence.' (Feist, 'The Evolved Fluid Specificity of Human Creative Talent', p. 69).

73 **Jeremy Bentham began studying Latin at age three:** Dinwiddy, *Bentham*, p. 11.

73 **John von Neumann could divide eight-digit numbers in his head by age six:** Myhrvold, 'John von Neumann'.

73 **Seattle's Adora Svitak began writing stories at age five and published her first book at age seven:** Bate, '"Dora the Explorer" shows pupils the way'.

73 **Ellen Winner responded in 2000 that 'Ericsson's research demonstrates the importance of hard work but does not rule out the role of innate ability . . . [We] conclude that intensive training is necessary for the acquisition of expertise, but *not* that it is sufficient.'**

—•» Winner also carefully reviewed now-known key ingredients of early achievement – motivation, independence, high expectations, and family nurturance – and, one by one, hypothesised that each could theoretically be consequences of innate giftedness rather than independent environmental ingredients:

> Gifted children have a deep intrinsic motivation to master the domain in which they have high ability, and are almost manic in their energy level . . . This intrinsic drive is part and parcel of an exceptional, inborn giftedness.
>
> Parents of gifted children grant their children more than the usual amount of independence. But we do not know whether granting independence leads to high achievement, or whether it is the recognition of the child's gift that leads to the granting of independence. It is also possible that gifted children are particularly strong willed and single minded and thus demand independence.
>
> Parents of gifted children typically have high expectations, and also model hard work and high achievement themselves. But it is logically possible that gifted children have simply inherited their gift from their parents, who also happen to be hardworking achievers.
>
> The families of gifted children are child-centred, meaning that family life is often totally focused on the child's needs. But the fact that parents spend a great deal of time with their gifted child does not mean that they create the gift. It is likely that parents first notice signs of exceptionality, and then respond by devoting themselves to the development of their child's extraordinary ability. (Winner, 'The origins and ends of giftedness'.)

While all these statements are logically plausible, they are each challenged by the evidence, by common sense, and by their own extreme unidirectionality. To declare with confidence that intrinsic motivation is inborn is to blatantly ignore early human psychology. While it's clear that biology contributes to personality, there's every evidence that it is not the sole determinant. To suggest that childhood independence could be caused wholly by the actions of a child is absurd. To suggest that parents' high expectations and modelling of hard work and high achievement could possibly have zero effect on a child because that child has simply inherited the 'gift' of motivation and talent from their parents is to embrace a genetic determinism even stronger than that of Galton. And finally, to say it is 'likely' that the child-centredness of families with precocious children begins wholly after the discovery of an exceptional ability is to ignore the variety of parenting styles the world around.

73 **'Necessary but not sufficient' became a common reaction to Ericsson as many professionals clung to the unsustainable notion of innate gifts:** For example, John Cloud, 'Is Genius Born or Can It Be Learned?' *Time*, February 13, 2009.

74 **We also know for sure that early musical exposure can work the same way.**

Abrams, Michael. 'The Biology of . . . Perfect Pitch: Can Your Child Learn Some of Mozart's Magic?' *Discover*, December 1, 2001.

Dalla Bella, Simone, Jean-François Giguère, and Isabelle Peretz. 'Singing proficiency in the general population'. *Journal of the Acoustical Society of America* 1212 (February 2007): 1182–89.

Deutsch, Diana. 'Tone Language Speakers Possess Absolute Pitch'. Presentation at the 138th meeting of the Acoustical Society of America, November 4, 1999.

Dingfelder, S. 'Most people show elements of absolute pitch'. *Monitor on Psychology* 36, no. 2 (February 2005): 33.

Kalmus, H., and D. B. Fry. 'On tune deafness (dysmelodia): frequency, development, genetics and musical background'. *Annals of Human Genetics* 43, no. 4 (May 1980): 369–82.

Lee, Karen. 'An Overview of Absolute Pitch'. Published online at https://web space.utexas.edu/kal463/www/abspitch.html, November 16, 2005.

74 **Imperceptibly, like water evaporating into a rain cloud, tiny events pave the way for development in one direction or another.**

→ The sudden emergence may sometimes appear to happen, but it doesn't really happen. 'We found no rigorous evidence for the sudden emergence of superior abilities in both prodigies and gifted students,' reports Ericsson. (Ericsson et al., 'Giftedness and evidence for reproducibly superior performance', p. 34.)

74 **For example, Winner points out that mathematically and musically 'gifted' individuals tend to use both lobes of the brain for tasks usually dominated by the left hemisphere in individuals with normal abilities.**

Winner's citations:

Gordon, H. W. 'Hemisphere asymmetry in the perception of musical chords'. *Cortex* 6 (1970): 387–98.

Gordon, H. W. 'Left-hemisphere dominance of rhythmic elements in dichotically presented melodies'. *Cortex* 14 (1978): 58–70.

Gordon, H. W. 'Degree of ear asymmetry for perception of dichotic chords and for illusory chord localization in musicians of different levels of competence'. *Journal of Experimental Psychology: Perception and Performance* 6 (1980): 516–27.

Hassler, M., and N. Birbaumer. 'Handedness, musical attributes, and dichaptic and dichotic performance in adolescents: a longitudinal study'. *Developmental Neuropsychology* 4, no. 2 (1988): 129–45.

O'Boyle, M. W., H. S. Gill, C. P. Benbow, and J. E. Alexander. 'Concurrent finger-tapping in mathematically gifted males: evidence for enhanced right hemisphere involvement during linguistic processing'. *Cortex* 30 (1994): 519–26.

74 **artists, inventors, and musicians tend to have a higher proportion of language disorders.**

Winner's citations:

Winner, E., and M. Casey. 'Cognitive Profiles of Artists'. In *Emerging Visions: Contemporary Approaches to the Aesthetic Process*, edited by G. Cupchik and J. Laszlo. Cambridge University Press, 1993.

Winner, E., M. Casey, D. DaSilva, and R. Hayes. 'Spatial abilities and reading deficits in visual art students'. *Empirical Studies of the Arts* 9, no. 1 (1991): 51–63.

Colangelo, N., S. Assouline, B. Kerr, R. Huesman, and D. Johnson. 'Mechanical Inventiveness: A Three-Phase Study'. In *The Origins and Development of High Ability*, edited by G. R. Bock and K. Ackrill. Wiley, 1993, pp. 160–74.

Hassler, M. 'Functional cerebral asymmetric and cognitive abilities in musicians, painters, and controls'. *Brain and Cognition* 13 (1990): 1–17.

74–75 **Consider that 'genetics' actually means 'genetic expression', and that the uterine environment and after-birth events are both highly developmental.**

—⊸ Which is not the same as saying 'under your control'.

75 **He is one of an estimated one hundred living prodigious savants who have both severe impairments and extraordinary abilities:** Treffert, 'Savant Syndrome'.

From the 'Savant Syndrome' FAQ page:

How common is savant syndrome?

Approximately one in ten (10 %) of persons with autistic disorder have some savant skills. In other forms of development disability, mental retardation or brain injury, savant skills occur in less than 1 % of such persons (approximately 1:2000 in persons with mental retardation). Since these other forms of mental disability are much more common than autistic disorder however, it turns out that approximately 50 % of persons with savant syndrome have autistic disorder, and the other 50 % have some other form of developmental disability, mental retardation or brain injury or disease. Thus not all savants are autistic, and not all autistic persons are savants.

What is the range of savant skills?

Savant skills exist over a spectrum of abilities. The most common savant abilities are called splinter skills. These include behaviours such as obsessive preoccupation with, and memorisation of, music and sports trivia, licence plate numbers, maps, historical facts, or obscure items such as vacuum cleaner motor sounds, for example. Talented savants are those persons in whom musical, artistic, mathematical or other special skills are more prominent and highly honed, usually within an area of single expertise, and are very conspicuous when viewed against their overall handicap. The term prodigious savant is reserved for those very rare persons in this already uncommon condition where the special skill or ability is so outstanding that it would be spectacular even if it were to occur in a non-handicapped person. There are probably fewer than 100 prodigious savants living worldwide at the present time who would meet this high threshold of special skill.

75 **The group also includes Daniel Tammet:** Treffert and Wallace, 'Islands of Genius'.

75 **He estimates that approximately one in ten persons with autism has some savant skills:** See excerpts from the 'Savant Syndrome' FAQ, above.

75 **The syndrome, he explains, occurs when the brain's left hemisphere is severely damaged, inviting the right hemisphere (which is responsible for things like music and art) to compensate heavily for the loss.**

Niki Denison writes:

> In trying to determine what causes savant syndrome, scientists turn to an increasing body of evidence that shows that when a particular part of the brain is thrown out of commission, another part attempts to compensate. Many have come to believe that in savant syndrome, the left hemisphere of the brain is damaged, so the brain adapts by drawing more heavily on the right hemisphere, which is responsible for creativity and skills in things like art and music. The left hemisphere, which is the home of language, comprehension, and logical, sequential thinking, is more vulnerable to harmful prenatal influences because it develops later and more slowly than the right hemisphere.
>
> One theory holds that an excess of circulating testosterone can impair left-hemisphere development, causing nerve cells to migrate to the right hemisphere and overdevelop that part of the brain. Because testosterone reaches very high levels in male foetuses, this could explain why savant syndrome is six times more common in boys than in girls. (Denison, 'The Rain Man in All of Us', p. 30.)

—◆ Kim Peek, the human calculator who inspired the Dustin Hoffman character in *Rain Man*, was missing the corpus callosum in his brain – the portion of the brain that allows the left and right sides of the brain to talk to each other easily.

76 **'In the case of the prodigious savant, it appears to me, there is a marvellous coalescence of idiosyncratic brain circuitry [combined with] obsessive traits of concentration & repetition and tremendous encouragement & reinforcement from family, caretakers and teachers. Does some of that same possibility, a little Rain Man as it were, perhaps reside within each of us? I think that it does':** Treffert, 'Is There a Little "Rain Man" in Each of Us?'

More from Treffert:

> The idea that some savant capabilities – a little Rain Man – might reside in each of us rises from several observations. First, there have been instances reported of previously non-disabled, 'normal' persons in whom some previously latent savant skills emerged following a head injury, a phenomenon called 'acquired' savant syndrome. Second, Dr. Bruce

> Miller's work, as described in detail elsewhere on this site, documents 12 cases of elderly persons, previously non-disabled, with no extraordinary savant skills, whose savant abilities newly emerged, sometimes at a prodigious level, after a particular type of dementia – fronto-temporal dementia – began and progressed. Thirdly, some procedures such as hypnosis or sodium amytal interviews in non-disabled persons, and brain surface electrode exploration during certain types of neurosurgical procedures, provide evidence that a huge reservoir of memories lies dormant, and non-accessed, in each of us. Fourth, the images and memories that surface, often to our surprise, during some dreams, also tap that huge store of buried memories beyond that available in our everyday waking state. Finally, often as we relax or 'tune out' other distractions, sometime after 'retirement' for example, some previously hidden, latent interests, talents or abilities quite suddenly, and surprisingly, emerge. Sometimes that emergence is actually a re-kindling of some earlier childhood abilities, such as art, for whatever reason set aside with maturation and 'growing up'. (Treffert, 'Savant Syndrome'.)

Diane Powell adds:

> Our model of savant abilities suggests that our brains operate at two levels, the quantum and the classical. These two levels are no more exclusionary than classical (or Newtonian) physics and quantum mechanics. One major difference between them is that the forces in classical physics operate locally, whereas forces in quantum physics operate nonlocally. Both types of forces operate in our brains, which is why our brains can process consciousness both locally and nonlocally. Some people have conditions such as autism that shift the balance between local and nonlocal processes by knocking out the functioning of the neocortex. The rest of us can decrease this classical dominance by such mind-quieting practices as meditation. Hence, as we become more consciously aware or awake, we use nonlocal processes more and more. Along the way, we will progressively see the world less abstractly. We will see it more as it really is. (Powell, 'We Are All Savants', p. 17.)

76 **'Apart from brain impairment and magnetic stimulation,' they wrote, 'savant-like skills might also be made accessible by altered states of perception or by EEG-assisted feedback. [Oliver] Sacks provides support for the former view. He produced camera-like precise drawings only when under the influence of amphetamines. Early (savant-like) cave art has been attributed to mescaline induced perceptual states':** Snyder,

Mulcahy, Taylor, Mitchell, Sachdev, and Gandevia, 'Savant-like skills exposed in normal people by suppressing the left fronto-temporal lobe', pp. 149–58.

Snyder's Citations

PERCEPTION

Snyder, A. W., and D. J. Mitchell. 'Is integer arithmetic fundamental to mental processing? The mind's secret arithmetic'. *Proceedings of the Royal Society of London. Series B, Containing Papers of a Biological Character* 266 (1999): 587–92.

EEG-ASSISTED FEEDBACK

Birbaumer, N. 'Rain Man's revelations'. *Nature* 399 (1999): 211–12.

OLIVER SACKS

Sacks, Oliver. 'The Mind's Eye'. *New Yorker*, July 28, 2003, pp. 48–59.

CAVE ART

Humphrey, N. 'Comments on shamanism and cognitive evolution'. *Cambridge Archaeological Journal* 12, no. 1 (2002): 91–94.

76–77 **It was his contention that the most successful children were endowed with elite genes propelling them to lifelong success. To prove this, he began tracking nearly fifteen hundred California schoolkids identified as 'exceptionally superior'.**

Ann Hulbert writes:

> Since Terman didn't have the resources to comprehensively test the more than a quarter-million students in the California school districts he was looking at, he enlisted teachers to help make the first cut. They supplied him with the kids they considered the best, a group unlikely to include 'some nerdy person in the corner mumbling to himself,' points out Dean Keith Simonton, a professor of psychology at the University of California, Davis, who specialises in the scientific study of historical genius. Testing this cohort – as well as other batches of bright children he rounded up earlier – Terman emerged with an overwhelmingly white and middle-class sample of roughly 1,500 students whose average age was 11 and whose I.Q.s ranged between 135 and 200, about the top 1 per cent. (The mean I.Q. in this group was 151, and 77 subjects tested at 170 or higher.) It is worth noting that his methods selected for a conscientious breed of parents as well, given that lengthy questionnaires

about their children were part of the drill. (Hulbert, 'The Prodigy Puzzle'.)

—⊸ The group was mostly middle class and mostly white; there were just two African Americans, which Terman took care to note 'are both *part* white . . . exact proportion of white blood is not known.' (Italics mine.) (Terman, *Genetic Studies of Genius: Volume I, Mental and Physical Traits of a Thousand Gifted Children*, p. 56.)

—⊸ In his first report, published in 1925, Terman tried to temper his expectations. 'To expect all or even a majority of the subjects to attain any considerable degree of eminence would be unwarranted,' he warned. But still, he could not contain his optimism: 'It is with the most distinguished 25 to 50 of [any average group of 5,000 adults] that our gifted boys could be most fairly compared a few decades hence.' (Terman, *Genetic Studies of Genius: Volume I, Mental and Physical Traits of a Thousand Gifted Children*, p. 641.)

77 **None went on to earn the Nobel Prize – as two children *rejected* from Terman's original group did.**

Ann Hulbert writes:

> In 1956, the year Terman died, a Nobel Prize was awarded to William Shockley, who as a California schoolboy didn't make the cut for the Termites but went on to help invent the transistor (and was later hailed as a catalyst in the creation of Silicon Valley, and also pilloried as a racist eugenicist). In 1968, another reject, Luis Alvarez, won the prize for his work in elementary particle physics. No Termite ever became a Nobel laureate, though some became well-published scientists and multiple patent holders. Alumni include journalists, poets and movie directors as well as professors, among whom psychologists have been particularly distinguished, perhaps not surprisingly. Terman, after all, pulled Stanford strings and did everything he could to help his protégés, who had been selected for what are often now called 'schoolhouse gifts' and had grown up as a self-identified group imbued, not least by him, with expectations of academically approved achievement.
>
> The fact that 'the group has produced no great musical composer,' as the study's authors wistfully noted, 'and no great creative artist' perhaps wasn't so surprising, either. (Hulbert, 'The Prodigy Puzzle'.)

> Holahan & Sears found that the 'Termites' in their seventies and eighties were no more successful in adulthood than if they had been randomly selected from the same socio-economic backgrounds – regardless of their IQ scores. This was somewhat mirrored in the findings of Subotnik, Kassan, Summers & Wasser (1993) who investigated a sample of 210 New York children selected for the Hunter College Elementary School by nomination and high-IQ scores (mean IQ 157). None had reached eminence by the ages of 40 to 50, nor were they any more successful than their socio-economic and IQ peers in spite of their tailor-made gifted education. (Freeman, 'Giftedness in the Long Term', pp. 384–403.)

77 **'One is left with the feeling that the above-180 IQ subjects were not as remarkable as might have been expected,' concluded Tufts's David Henry Feldman in a 1984 retrospective of the long study. 'There is the disappointing sense that they might have done more with their lives.'**

The entire text of the quote:

> On the whole, one is left with the feelings that the above-180 IQ subjects were not as remarkable as might have been expected. Without question they have done better than the general population in most major categories and there is some evidence (although not a great deal) that they were more successful in their careers than the 150 IQ group. But, when we recall Terman's early optimism about his subjects' potential, and the words of Hollingworth (1942) that 'the children who test at above 180 IQ constitute the "top" among college graduates,' there is the disappointing sense that they might have done more with their lives. (Feldman, 'A follow-up of subjects scoring above 180 IQ in Terman's genetic studies of genius', pp. 518–23.)

Ann Hulbert adds:

> Focusing on a small cohort of children with I.Q.s above 180, [Leta] Hollingworth's case studies couldn't supply clear-cut evidence that a high-testing childhood was a precursor of later extraordinariness. (Hulbert, 'The Prodigy Puzzle'.)

77 **'Most gifted children, even most child prodigies, do not go on to become adult creators':** Winner, 'The origins and ends of giftedness', pp. 159–69.

Ericsson strongly affirms this point:

Notably, there are only comparatively few prodigies, such as Mozart, Picasso, and Yehudi Menuhin, who continued their success into adulthood – most prodigies do not live up to expectations (Bamberger, 1986; Barlow, 1952; Freeman, 2000; Goldsmith, 2000). (Ericsson, Roring, Nandagopal, 'Giftedness and evidence for reproducibly superior performance: an account based on the expert performance framework', pp. 3–56.)

Ericsson's Citations

Bamberger, J. 'Growing Up Prodigies: The Mid-life Crisis'. In *Developmental Approaches to Giftedness and Creativity*, edited by D. H. Feldman. Jossey-Bass, 1986, pp. 61–67.

Barlow, F. *Mental Prodigies*. Greenwood Press, 1952.

Freeman, J. 'Teaching for Talent: Lessons from the Research'. In *Developing Talent Across the Lifespan*, edited by C. F. M. Lieshout and P. G. Heymans. Psychology Press, 2000, pp. 231–48.

Goldsmith, L. T. 'Tracking Trajectories of Talent: Child Prodigies'. In *Talents Unfolding*, edited by R. C. Friedman and B. M. Shore. American Psychological Association, 2000, pp. 89–118.

Middlesex University's Joan Freeman adds:

Trost (1993) calculated that less than half of 'what makes excellence' can be accounted for by measurements and observations in childhood. The key to success, he said, lies in the individual's dedication. (Freeman, 'Families, the essential context for gifts and talents', pp. 573–85; Trost, 'Prediction of Excellence in School, University and Work', pp. 325–36.)

77–78 **'Technical perfection wins the prodigy adoration, but if the prodigy does not eventually go beyond this, he or she sinks into oblivion.'**

More from Ellen Winner:

A creator is someone who changes a domain. Personality and will are crucial factors in becoming an innovator or revolutioniser of a domain. Creators have a desire to shake things up. They are restless, rebellious, and dissatisfied with the status quo. They are courageous and independent. They are able to manage multiple related projects at the same time, engaging in what Gruber calls a 'network of enterprise'. For these two reasons, we should never expect a prodigy to go on to become a creator. The ones who do make this transition are the exceptions, not the rule. (Winner, 'The origins and ends of giftedness', pp. 159–69.)

Joan Freeman writes:

> Subotnik, Kassan, Summers & Wasser (1993) have shown that giftedness may take many different forms; it may appear in quite unexpected situations and at different points during a lifetime. It is not always possible to identify future gifts. (Freeman, 'Giftedness in the long term', pp. 384–403.)

—•⇛ With all due respect to Professor Freeman, isn't the effort to 'identify future gifts' a slightly crazy way of discussing future achievement? If we step away from the 'giftedness' paradigm and simply regard achievements as achievements, the same research cited above would be restated simply: *Adults with undistinguished backgrounds and childhoods often turn out to be high achievers, and those achievements can happen at various points in their lives.*

78 **'Prodigies [can] become frozen into expertise,' says Ellen Winner. 'This is particularly a problem for those whose work has become public and has won them acclaim, such as musical performers, painters, or children who have been publicised as "whiz kids" . . . It is difficult to break away from [technical] expertise and take the kinds of risks required to be creative.'**

Ellen Winner on 'when giftedness ends':

> One non-inevitable reason that prodigies may fail to make the transition is that they have become frozen into expertise. This is particularly a problem for those whose work has become public and has won them acclaim, such as musical performers, painters, or children who have been publicised as 'whiz kids'. Expertise is what has won them fame and adoration as child prodigies. It is difficult to break away from expertise and take the kinds of risks required to be creative. A second non-inevitable reason is that some with the potential to make the transition do not do so because they have been pushed so hard by their parents and teachers and managers that they lose their intrinsic motivation. At adolescence they begin to ask, 'Who am I doing this for?' And if the answer is that they are doing this for a parent or a teacher but not for themselves, they may decide that they do not want to do this anymore. And so they drop out. The case of William James Sidis, a math prodigy pushed relentlessly by his father, *is one such case among many.* (Italics mine.) (Winner, 'The origins and ends of giftedness', pp. 159–69.)

Ann Hulbert writes:

> For at least a quarter century now, there has been 'a benevolent conspiracy' among influential musical figures to fend off burnout by trying to foster 'a more humanistic, nonexploitative approach to the development of talent,' as the writer Marie Winn put it in a *New York Times Magazine* article in 1979. What a researcher named Jeanne Bamberger has termed a 'midlife' crisis seems to occur for prodigious young musicians: a transitional period of cognitive and emotional maturation during which only some performers manage to move beyond intuitive imitation to a more reflective sense of direction. Parents must carve out space for precocious players to 'have a childhood . . . an adolescence,' according to influential figures like Itzhak Perlman; resist the pressure, they urge, to 'get management' and a packed schedule of practice and performance. (Hulbert, 'The Prodigy Puzzle'.)

78 What was the true source of Yo-Yo's uncanny ability? In her memoir, his mother chalks it up to genetics – but then she details how, from the very moment of his birth, Yo-Yo was exposed to music in the most profound and exquisite way.

—⊸ Marina Ma calls it a genetic gift in her book, but to my eyes, this comment is obviously a combination of her cultural humility and her being a little too close to detect the forest of details that spurred Yo-Yo on.

79 'From the cradle, Yo-Yo was surrounded by a world of music,' his mother recalls. 'He heard hundreds of classical selections on records, or played by his father or his sister. Bach and Mozart were engraved on his mind.'

—⊸ And let's not forget what can happen *before* birth. Here is Giselle E. Whitwell's thorough review of the profound impact that sound can have on a foetus in utero:

> Verny and others have noted that babies have a preference for stories, rhymes, and poems first heard in the womb. When the mother reads out loud, the sound is received by her baby in part via bone conduction. Dr. Henry Truby, Emeritus Professor of Pediatrics and Linguistics at the University of Miami, points out that after the sixth month, the foetus moves in rhythm to the mother's speech and that spectrographs of the first cry of an abortus at 28 weeks could be matched with his mother's. The elements of music, namely tonal pitch, timbre, intensity and rhythm, are also elements used in speaking a language. For this reason, music

prepares the ear, body and brain to listen to, integrate and produce language sounds. Music can thus be considered a pre-linguistic language which is nourishing and stimulating to the whole human being, affecting body, emotions, intellect, and developing an internal sense of beauty, sustaining and awakening the qualities in us that are wordless and otherwise inexpressible. The research of Polverini-Rey (1992) seems to indicate that prenates exposed to lullabies in utero were calmed by the stimulus. The famous British violinist Yehudi Menuhin believes that his own musical talent was partly due to the fact that his parents were always singing and playing music before he was born.

The ear first appears in the 3rd week of gestation and it becomes functional by the 16th week. The foetus begins active listening by the 24th week. We know from ultrasound observations that the foetus hears and responds to a sound pulse starting about 16 weeks of age; this is even before the ear construction is complete. The cochlear structures of the ear appear to function by the 20th week and mature synapses have been found between the 24th and 28th weeks. For this reason most formal programmes of prenatal stimulation are usually designed to begin during the third trimester. The sense of hearing is probably the most developed of all the senses before birth. Four-month-old foetuses can respond in very specific ways to sound; if exposed to loud music, their heart beat will accelerate. A Japanese study of pregnant women living near the Osaka airport had smaller babies and an inflated incidence of prematurity – arguably related to the environment of incessant loud noise. Chronic noise can also be associated with birth defects. I recently received a report from a mother who was in her 7th month of pregnancy when she visited the zoo. In the lion's enclosure, the animals were in process of being fed. The roar of one lion would set off another lion and the sound was so intense she had to leave the scene as the foetus reacted with a strong kick and left her feeling ill. Many years later, when the child was 7 years of age, it was found that he had a hearing deficiency in the lower-middle range. This child also reacts with fear when viewing TV programmes of lions and related animals. There are numerous reports about mothers having to leave war movies and concerts because the auditory stimulus caused the fetus to become hyperactive.

Chamberlain (1998), using Howard Gardner's concept of multiple intelligences, has presented evidence for musical intelligence before birth. Peter Hepper (1991) discovered that prenates exposed to TV soap opera music during pregnancy responded with focused and rapt attention to this music after birth – evidence of long-term memory. On hearing the music after birth, these newborns had a significant decrease in heart rate and movements, and shifted into a more alert state. Likewise, Shetler (1989) reported that 33% of foetal subjects in his study demonstrated contrasting reactions to tempo variations between faster and slower selections of music. This may be the earliest and most primitive musical response in utero. The pioneering New Zealand fetologist, William Liley, found that from at least 25 weeks on, the unborn child would jump in rhythm with the timpanist's contribution to an orchestral performance. The research of Michele Clements (1977) in a London maternity hospital found that four to five month foetuses were soothed by Vivaldi and Mozart but disturbed by loud passages of Beethoven, Brahms and Rock. Newborns have shown a preference for a melody their mother sang in utero rather than a new song sung by their mother. Babies during the third trimester in utero respond to vibroacoustic as well as air-coupled acoustic sounds, indicative of functional hearing. A study by Gelman et al. (1982) determined that a 2000 Hz. stimulus elicited a significant increase in foetal movements, a finding which supported the earlier study by Johnsson et al. (1964). From 26 weeks to term, foetuses have shown fetal heart accelerations in response to vibroacoustic stimuli. Consistent startle responses to vibroacoustic stimuli were also recorded during this period of development. Behavioural reactions included arm movements, leg extensions, and head aversions. Yawning activity was observed after the conclusion of stimuli. Research by Luz et al. (1980 and 1985) has found that the normal foetus responds to external acoustic stimulation during labour in childbirth. These included startle responses to the onset of a brief stimulus. New evidence of cognitive development in the prenatal era is presented by William Sallenbach (1994) who made in-depth and systematic observations of his own daughter's behaviour from weeks 32 to 34 in utero. (The full report of his findings is available on this Web site in Life Before Birth/Early Parenting.) Until recently, most research on early learning processes has been in the area of habituation, conditioning or imprinting sequences. However, Sallenbach observed that in the last trimester of pregnancy, the prenate's learning state shows movement from abstraction and generalisation to one of increased specificity and differentiation. During a bonding session using music, the prenate was observed moving her hands gently. In a special musical arrangement, where dissonance was included, the subject's reactions were more rhythmic with rolling movements. Similarly, in prenatal music classes, Sister Lorna Zemke has found that the foetus will respond rhythmically to rhythms tapped on the mother's belly. (Whitwell, 'The Importance of Prenatal Sound and Music'.)

79 'Melodic "calculation centres" in the dorsal temporal lobes appear to be paying attention to interval size and distances between pitches as we listen to music': Levitin, *This Is Your Brain on Music*, p. 160; see also Münte, Altenmüller, Jäncke, et al., 'The musician's brain as a model of neuroplasticity', pp. 473–78, and Weinberger, 'Music and the Brain', 88–95.

79–80 Levitin also concurs with University of California, San Diego's Diana Deutsch and others in deducing that every human being is likely born with the capacity for absolute pitch, but that it gets activated only in those who are exposed to enough tonal imprinting at a very early age.

—⊸ Glenn Gould had it – so did Beethoven, Bach, Mozart, Horowitz, and Sinatra. On the surface, absolute pitch seems like the province of musical geniuses – the exotic gift that they have and we don't. But the truth about absolute pitch – and the opposite phenomenon of so-called tone deafness – is much more interesting, and helps us understand what 'musical talent' really is and isn't.

What is absolute pitch?

Absolute pitch (AP) is the ability to produce and identify a certain musical tone without any reference tone. A person with AP is able to hum middle C or any other note on request, without any prompting from a song or an instrument.

How common is AP?

In strict definitional terms, AP is pretty rare – somewhere between 1 in 10,000 and 1 in 2,000 in the general population. But the rare part is the note naming, not the note reproducing. Many studies have now shown that most people can sing a familiar song in the right key without being given a reference tone and that virtually everyone who speaks a tonal language such as Mandarin can remember and recall specific pitches. What few people possess is the specific trained ability to link that tone to a named note.

'Our studies tie right in with the idea that we all have this latent absolute pitch ability, but we can't get fully bloomed absolute pitch without early childhood training,' says Shepherd College's Laura Bischoff.

'The real puzzle about perfect pitch is not why so few people possess it but rather why most people do not,' UC San Diego's Diana Deutsch says. 'Everyone has an implicit form of perfect pitch, even though we aren't all able to put a label to notes. They can recognise the note but can't label it. What's learned as a child is the ability to label.'

Also, contrary to public assumption, AP is not an all-or-nothing skill.

Many have AP in varying degrees, explain Bischoff and the University of Rochester's Elizabeth West Marvin.

Is AP a critical ingredient in musical talent?

No. While AP can sometimes be a useful tool for musicians, it is far from essential in helping musicians build the necessary skills or in expressing themselves magnificently. AP is more common among professional musicians than nonmusicians, but research shows very clearly that this is not cause and effect. Rather, the correlation exists because both are so frequently a product of early (prior to age six) musical training.

Neither Wagner nor Stravinsky had AP, to name just two. McGill University's Daniel Levitin (author of *This Is Your Brain on Music*) does not think AP helps musicians much. What musicians thrive on and must develop to a fine degree, he points out, is relative pitch – the ability to distinguish between tones. Such relative pitch is available to almost everyone, to be developed to whatever individual degree desired.

'The average person is able to carry a tune almost as proficiently as professional singers. This result is consistent with the idea that singing is a basic skill that develops in the majority of individuals, enabling them to engage in musical activities. In short, singing appears to be as natural as speaking.' (Dalla Bella et al., 2007.)

What about 'tone-deaf' people who can't carry a tune?

So-called tone deafness is a little-studied and much-misunderstood subject now getting closer attention. Four per cent of the general population has tone deafness (Kalmus and Fry, 1980), which until recently was thought to be mainly a perceptual deficit – affected individuals supposedly could not hear the difference in tones; they did not have and could not develop relative pitch, and therefore could not appreciate or produce music.

New evidence has forced an entirely new conclusion. Studies now show that virtually everyone can distinguish tonal differences and appreciate music (Dalla Bella et al., 2007). And while a tiny percentage of people truly cannot hear tonal differences due to some specific brain damage, 'present findings suggest that tone-deafness may emerge as a pure output disorder . . . that poor singing may occur in the presence of normal perception. This possibility finds support in a recent study conducted with poor singers who exhibited pitch production deficits but normal pitch discrimination' (Bradshaw & McHenry, 2005).

In other words, the vast majority of people who call themselves tone-deaf (or who are mocked as such by friends and spouses) actually hear and perceive music perfectly well and simply have a problem generating with their vocal chords the tones they hear in their brain.

Sources cited in the text above:

Dickinson, Amy. 'Little Musicians'. *Time,* December 13, 1999.

Brown, Kathryn. 'Striking the Right Note'. *New Scientist,* December 4, 1999.

Dingfelder, S. 'Most people show elements of absolute pitch'. *Monitor on Psychology* 36, no. 2 (February 2005): 33.

Abrams, Michael. 'The Biology of . . . Perfect Pitch: Can Your Child Learn Some of Mozart's Magic?' *Discover,* December 1, 2001.

Deutsch, Diana. 'Tone Language Speakers Possess Absolute Pitch'. Presentation at the 138th meeting of the Acoustical Society of America, November 4, 1999.

Lee, Karen. 'An Overview of Absolute Pitch'. Published online at https://web space.utexas.edu/kal463/www/abspitch.html, November 16, 2005.

Dalla Bella, Simone, Jean-François Giguère, and Isabelle Peretz. 'Singing proficiency in the general population'. *Journal of the Acoustical Society of America* 1212 (February 2007): 1182–89.

Kalmus, H., and D. B. Fry. 'On tune deafness (dysmelodia): frequency, development, genetics and musical background'. *Annals of Human Genetics* 43, no. 4 (May 1980): 369–82.

Bradshaw, E., and M. A. McHenry. 'Pitch discrimination and pitch matching abilities of adults who sing inaccurately'. *Journal of Voice* 19, no. 3 (September 2005): 431–39.

80 **Yo-Yo worshipped his sister and father and desperately wanted to impress both:** Ma, *My Son, Yo-Yo*, p. 27.

80 **Ellen Winner calls it 'the rage to master', a fervent, never-let-go willfulness and focus that drives a child into an early version of Ericsson's deliberate practice.**

Winner writes:

> Gifted children have a deep intrinsic motivation to master the domain in which they have high ability, and are almost manic in their energy level. Often one cannot tear these children away, whether from an instrument, a computer, a sketch pad, a math book. They have a

powerful interest in the domain in which they have high ability, and they can focus so intently on work in this domain that they lose sense of the outside world. These children combine an obsessive interest with an ability to learn easily in a given domain. Unless social and emotional factors interfere, this combination leads to high achievement. This intrinsic drive is part and parcel of an exceptional, inborn giftedness. (Winner, 'The origins and ends of giftedness', pp. 159–69.)

—⊸ Winner insists that this rage to master is innate, but only because she can't deduce an external cause. There's no proof offered at all, other than it just seems to appear in kids' lives (albeit only in child-centred families where parents are semi-obsessed themselves with their kids' skills). The obvious possibility that rage to master is a psychological mechanism forming out of some family/social/cultural dynamic does not seem even to be considered. This is a shame, because Winner seems to have a keen understanding of so many other facets of giftedness, including the psychodynamics of a gifted child growing into adolescence and struggling to maintain that intrinsic motivation.

For much more on this, see endnote on p. 262: 'The brain circuits that modulate a person's level of persistence are plastic – they *can* be altered'.

80 As a general rule, high achievers have exceptional drive.

—⊸ Joan Freeman has done much important writing on this subject. Here, she relates a raft of research that points to the importance of attitude, as opposed to early success:

> In the Scottish study, childhood intelligence was not always related to how people perceived their success in life. The most reliable predictor in their early years was found to be positive self-esteem, and the most useful tools for actually climbing the career ladder were optimism and pugnacity, similar to what Moon (2002) calls Personal Talent *which she describes as teachable.* Indeed, Trost (2000), investigating prediction of giftedness in adult life, calculated that less than half of 'what makes excellence' can be accounted for by measurements and observations in childhood: for intelligence not more than 30%. The key to success, he wrote, lies in the individual's dedication. Others have suggested optimism as the key. (Italics mine.) (Freeman, 'Giftedness in the long term', pp. 384–403.)

81 Michael Jordan always seemed to hate losing (an everyday experience while growing up with his brother Larry).

—⊸ His friend Roy Smith reports that in junior high school if you played H.O.R.S.E. with Jordan and won, that simply meant you'd play another game, and then another and then another, until you lost. Then you could go home. (Halberstam, *Playing for Keeps*, p. 21.)

81 'There were nine players on the court just coasting,' Coley recalls: Halberstam, *Playing for Keeps*, p. 22.

81 'Even in pickup games,' writes Halberstam, 'he had become unusually purposeful.'

—⊸ The special signature of Jordan's psychology, writes David Halberstam, was that he could turn anything into a personal slight that demanded personal revenge. (Halberstam, *Playing for Keeps*, p. 98.)

83 Other Dweck experiments pointed in the same direction, demonstrating irrefutably that people who believe in inborn intelligence and talents are less intellectually adventurous and less successful in school. By contrast, people with an 'incremental' theory of intelligence – believing that intelligence is malleable and can be increased through effort – are much more intellectually ambitious and successful.

—⊸ The researchers first measured the subjects' beliefs and then tracked them for two years through seventh and eighth grades (age 12–14). Blackwell, Trzesniewski, and Dweck write:

> Nearly two years later, students who endorsed a strong incremental theory of intelligence at the beginning of junior high school were outperforming those who held more of an entity theory in the key subject of mathematics, controlling for prior achievement. Moreover, their motivational patterns mediated this relation such that students with an incremental orientation had more positive motivational beliefs, which in turn were related to increasing grades . . .
>
> This research confirms that adolescents who endorse more of an incremental theory of malleable intelligence also endorse stronger learning goals, hold more positive beliefs about effort, and make fewer ability-based, 'helpless' attributions, with the result that they choose more positive, effort-based strategies in response to failure, boosting mathematics achievement over the junior high school transition. Furthermore, this motivational framework at the beginning of junior high school was related to the trajectories of students' math achievement over the two years of junior high school: students who endorsed a more

incremental theory framework increased in math grades relative to those who endorsed a more entity theory framework, showing that the impact of this initial framework remained predictive over time . . . Within a single semester, the incremental theory intervention appears to have succeeded in halting the decline in mathematics achievement.

Further, these findings support the idea that the diverging achievement patterns emerge only during a challenging transition. Prior to junior high school, students who endorsed more of an entity theory seemed to be doing fine. As noted in previous research, motivational beliefs may not have an effect until challenge is present and success is difficult. *Thus, in a supportive, less failure-prone environment such as elementary school, vulnerable students may be buffered against the consequences of a belief in fixed intelligence. However, when they encounter the challenges of middle school, these students are less equipped to surmount them.* (Italics mine.) (Blackwell, Trzesniewski, and Dweck, 'Implicit theories of intelligence predict achievement across an adolescent transition', pp. 246–63; see also Bronson, 'How Not to Talk to Your Kids'.)

83 **Regardless of whether a child seems to be exceptional, mediocre, or even awful at any particular skill at a particular point in time, the potential exists for that person to develop into a high-achieving adult.**
San Jose State University's Gregory Feist writes:

> It is important to point out, just as is true of mathematical precocity and prodigiousness, early childhood talent in music by no means is a necessary or a sufficient condition for adult creative achievement. It is often the case that the musically most-accomplished adults do not begin to set themselves apart in any significant way until middle adolescence, and even here there are hundreds if not thousands of similarly talented musicians. It is also true that being a musical prodigy or even being precocious does not guarantee or even predict to a high degree adult creative achievement. (Feist, 'The Evolved Fluid Specificity of Human Creative Talent', p. 69.)

83 **Because talent is a function of acquired skills rather than innate ability, adult achievement depends completely on long-term attitude and resources and process rather than any particular age-based talent quotient.**

—→ This does not, of course, mean that everything is within our control, as is discussed in chapter 7.

CHAPTER 6: CAN WHITE MEN JUMP? ETHNICITY, GENES, CULTURE, AND SUCCESS

PRIMARY SOURCES

Entine, Jon. *Taboo: Why Black Athletes Dominate Sports and Why We Are Afraid to Talk About It*. Public Affairs, 2000.

Noakes, Timothy David. 'Improving Athletic Performance or Promoting Health Through Physical Activity'. World Congress on Medicine and Health, July 21–August 31, 2000.

CHAPTER NOTES

84 **At the 2008 Summer Olympics in Beijing.**
2008 Olympic track and field results for Jamaican medalists:

Men's 100-metre final: Usain Bolt (gold) 9.69 seconds
Men's 200-metre final: Usain Bolt (gold) 19.30 seconds
Women's 100-metre final: Shelly-Ann Fraser (gold) 10.78 seconds, Kerron Stewart (silver) 10.98 seconds, Sherone Simpson (silver) 10.98 seconds
Women's 200-metre final: Veronica Campbell-Brown (gold) 21.74 seconds, Kerron Stewart (bronze) 22.00 seconds
Women's 400-metre final: Shericka Williams (silver) 49.69 seconds
Women's 400-metre hurdles final: Melaine Walker (gold) 52.64 seconds
Men's 4 x 100-metre relay: Nesta Carter, Michael Frater, Usain Bolt, Asafa Powell (gold) 37.10 seconds
Women's 4 x 400-metre relay: Shericka Williams, Shereefa Lloyd, Rosemarie Whyte, Novelene Williams (bronze) 3 minutes 20.40 seconds

Total Jamaican medals: six gold, three silver, two bronze
JamaicaOlympicGlory.com Web site.

84 **'They brought their A game':** Phillips, 'Jamaica Gold Rush Rolls On, US Woe in Sprint Relays'.

84 **Within hours, geneticists and science journalists rushed in with reports of a 'secret weapon':** Fest, ' "Actinen A", Jamaica's secret weapon'; see also Olympics Diary, 'Jamaicans built to beat the rest'.

85 **'no clear relationship between the frequency of this variant in a population and its capacity to produce sprinting superstars':** MacArthur, 'The Gene for Jamaican Sprinting Success? No, Not Really'.

85 **This is the same question people asked about champion long-distance runners from Finland in the 1920s and about great Jewish basketball players from the ghettos of Philadelphia and New York in the 1930s. Today, we wonder how tiny South Korea turns out as many great female golfers as the United States – and how the Dominican Republic has become a factory for male baseball players:** Bale, *Sports Geography*, pp. 60, 72.

—◦➾ To be clear, 'great Jewish basketball players' is not a joke. Jon Entine notes the success of Jewish players in the 1930s:

'The reason, I suspect, that basketball appeals to the Hebrew with his Oriental background,' wrote Paul Gallico, sports editor of the *New York Daily News* and one of the premier sportswriters of the 1930s, 'is that the game places a premium on an alert, scheming mind, flashy trickiness, artful dodging and general smart aleckness.' Writers opined that Jews had an advantage in basketball because short men have better balance and more foot speed. They were also thought to have sharper eyes, which of course cut against the stereotype that Jewish men were myopic and had to wear glasses. (Entine, 'Jewish hoop dreams'.)

85 **'sports geography' has developed over the years to help understand it.**

Some prominent sports geographers: John Bale, Joseph Maguire, Harold McConnell, Carl F. Ojala, Michael T. Gadwood, John F. Rooney, G. A. Wiggins, and P. T. Soule.

85 **In his book *Taboo: Why Black Athletes Dominate Sports and Why We're Afraid to Talk About It*, journalist Jon Entine insists that today's phenomenal black athletes – Jamaican sprinters, Kenyan marathoners, African-American basketball players, etc. – are propelled by 'high performance genes' inherited from their West and East African ancestors.**

—◦➾ Blacks descended from West Africans, Entine explains, are endowed with shorter trunks and smaller lungs, longer arms and legs, narrower hips, heavier bones, more muscle all around, less subcutaneous fat, a higher centre of gravity, a higher bone density, and a much higher proportion of 'fast-twitch' muscle fibres – all key ingredients for strength-based, short-burst aerobic sports.

Meanwhile, three thousand miles across the continent, Entine explains, the same evolutionary forces have bestowed *East* Africans with a very different set of 'high performance genes'. This lucky breed has smaller physiques, narrow shoulders, lean legs, much less muscle mass, and a higher proportion of 'slow-twitch' muscles, rendering them ideal endurance athletes: marathon runners, cyclists, swimmers, etc.:

> Relative advantages in these physiological and biomechanical characteristics are a gold mine for athletes who compete in such anaerobic activities as football, basketball, and sprinting, sports in which West African blacks clearly excel . . . East Africa produces the world's best aerobic athletes because of a variety of bio-physiological attributes. (Entine, *Taboo*, p. 269.)

85–86 'White athletes appear to have a physique between central West Africans and East Africans,' Entine writes. 'They have more endurance but less explosive running and jumping ability than West Africans; they tend to be quicker than East Africans but have less endurance.'

—⊸⪢ Physiologically, Entine tells us, they're stuck somewhere in the middle, leaving them without particular advantages in either short-burst or endurance sports. (Entine, *Taboo*, p. 269.)

86 In his own book, Entine quotes geneticist Claude Bouchard: 'The key point is that these biological characteristics *are not unique* to either West or East African blacks. These characteristics are seen in all populations, including whites.'

Bouchard continues: 'However, based on the limited number of studies available, there seem to be more African Blacks with such characteristics than there are in other populations.' (Entine, *Taboo*, p. 261.)

Entine also quotes others making the same point: 'An average advantage, yes, but that says nothing about any individual competitor,' says Lindsay Carter. 'You've got to be very careful generalising,' warns Michigan State's Robert Malina. (Entine, *Taboo*.)

86 Entine also acknowledges that we haven't in fact found the actual genes he's alluding to. 'These genes will likely be identified early in the [twenty-first century],' he predicts.

Still, he contends, these as-yet-unfound genes are critical. 'All the hard work in the world will go for naught if the roulette wheel of genetics doesn't land on your number.' (Entine, *Taboo*, p. 270.)

86 **'It's pointless for me to run on the pro circuit,' complained American 10,000-metre champion Mike Mykytok:** Bloom, 'Kenyan Runners in the U.S. Find Bitter Taste of Success'.

86 **'The better a young man was at raiding [cattle]':** Manners, 'Kenya's running tribe'.

87 **He wasn't the most precocious or 'natural' athlete:** Bale, comment on The Sports Factor radio show, February 28, 1997.

87 **'I used to run from the farm to school and back,' he recalled:** Entine, *Taboo*, p. 51.

88 **In the decades that followed, the long-standing but profitless Kalenjin running tradition became a well-oiled economic-athletic engine.**

Alexander Wolff writes on the Kenyan running 'miracle':

> Salazar ticks off the ironic circumstances that seem to cast the U.S. as a Third World country in distance running: 'As big as we are, we have fewer people to draw on. In Kenya there are probably a million schoolboys 10 to 17 years old who run 10 to 12 miles a day . . . The average Kenyan 18-year-old has run 15,000 to 18,000 more miles in his life than the average American – and a lot of that's at altitude. They're motivated because running is a way out. Plus they don't have a lot of other sports for kids to be drawn into. Numbers are what this is all about. In Kenya there are maybe 100 runners who have hit 2:11 in the marathon – and in the U.S. maybe five . . .
>
> With those figures, coaches in Kenya can train their athletes to the outer limits of endurance – up to 150 miles a week – without worrying that their pool of talent will be meaningfully depleted. Even if four out of every five runners break down, the fifth will convert that training into performance . . . (Wolff, 'No Finish Line'.)

Commenting on this Wolff article on his Web site, Malcolm Gladwell writes:

> We've always known that running is culturally important in Kenya, in a way it isn't anywhere else in the world. But these are staggering numbers. A *million* 10 to 17 year olds running *10 to 12* miles a day? I'm

guessing the United States doesn't have more than 5,000 or so boys in that age bracket logging that kind of mileage. [Seventy] miles a week is an enormous amount of running – even for an adult. I ran middle distance at a nationally competitive level as a teenager, and never got close to 70 miles a week.

I know this isn't going to put the genetic argument about Kenyan running dominance to rest. But maybe it should. It's a far more parsimonious explanation. No one ever claims that Canadians are genetically superior to everyone else when it comes to hockey, or that Dominicans have a genetic advantage when it comes to baseball. We all accept the fact that those two countries succeed at those sports because they draw their elite talent from a developmental pool that is simply larger – in relative and in some cases absolute terms – [than] other nations. [It's] a numbers game. If Kenya really has a million kids, doing that kind of mileage, then we scarcely need any other explanation for their success.

Here's the appropriate thought experiment. Imagine that every year 50 per cent of all American 10 year old boys were shipped to Boulder, Colorado, where they ran 50 to 70 miles a week at altitude for the next seven years. Would the United States regain control of international middle and long distance running? (Gladwell, 'Kenyan Runners'.)

88 **High-altitude training and mild year-round climate are critical:**

Sir Roger Bannister's statement, that it would take a lifetime for an athlete born at sea level to adapt for maximum exercise at medium altitude, was proved correct. (Noakes, 'Improving Athletic Performance or Promoting Health through Physical Activity'.)

88 **In testing, psychologists discovered a particularly strong cultural 'achievement orientation':** Hamilton, 'East African running dominance', pp. 391–94.

—•➾ Much research has been conducted on individuals 'high in achievement motivation' (HAMs). In 1938, H. A. Murray defined HAMs as those who seek challenge, desire to attain competence, and strive to outdo others.

Psychologists John M. Tauer and Judith M. Harackiewicz write:

Our results provide strong evidence that the effects of competition on intrinsic motivation are moderated by achievement orientation, even when feedback is not provided. Our findings converge with those of Study 1 to suggest that HAMs and LAMs [individuals low in achievement motivation] respond to competition very differently . . .

Clearly, positive feedback is not the reason HAMs enjoy activities in competition. In Study 1, HAMs enjoyed Boggle more in competition than LAMs, even when they received negative feedback. In Study 2, we observed similar reactions in the absence of any outcome feedback. Taken together, these results clearly demonstrate that the differential effects of competition are due to the competitive context established at the beginning of competition . . .

The results of this study are therefore consistent with Joe Paterno's claim that competition can be enjoyable regardless of whether one wins or loses. (Tauer and Harackiewicz, 'Winning isn't everything', pp. 209–38.)

88 ***How can the rest of the world defuse Kenyan running superiority? Answer: Buy them school buses***: Wolff, 'No Finish Line'.

89 **'coaches in Kenya can train their athletes to the outer limits of endurance'**: Wolff, 'No Finish Line'.

89 **And what of genetics? Are Kenyans the possessors of rare endurance genes, as some insist? No one can yet know for sure, but the new understanding of GxE and some emergent truths in genetic testing strongly suggest otherwise.**

Some pertinent comments on this from T. D. Brutsaert and E. J. Parra:

> First, the cumulative evidence, going back more than one century, is all but overwhelming in support of the general idea that genes are responsible for some of the variation in human athletic performance.
>
> The second point is that despite the obvious role of genetics in human physical performance, there is little unequivocal evidence in support of a specific genetic variant with a major gene effect on a relevant performance phenotype.
>
> Much like the complex genetic and environmental etiology of chronic disease, athletes likely emerge on a predisposing and favourable genetic background where individual alleles are both common and have only modest effects.
>
> The challenge for exercise science is to incorporate an even broader concept of the environment to include environmental influences that act, not just over the short term, but during critical periods of

development including prenatal life, early childhood, and adolescence. (Brutsaert and Parra, 'What makes a champion?' p. 110.)

89 Skin colour is a great deceiver; actual genetic differences between ethnic and geographic groups are very, very limited.

According to researchers at the National Human Genome Research Institute:

> A prominent exception to the common distribution of physical characteristics within and among groups is skin colour. Approximately 10 % of the variance in skin colour occurs within groups, and ~90 % occurs between groups (Relethford 2002). This distribution of skin colour and its geographic patterning – with people whose ancestors lived predominantly near the equator having darker skin than those with ancestors who lived predominantly in higher latitudes – indicate that this attribute has been under strong selective pressure. (Berg et al., 'The use of racial, ethnic, and ancestral categories in human genetics research', pp. 519–32.)

89 All human beings are descended from the same African ancestors.

Kate Berg writes:

> The existing fossil evidence suggests that anatomically modern humans evolved in Africa, within the last ~200,000 years, from a pre-existing population of humans (Klein 1999). Although it is not easy to define 'anatomically modern' in a way that encompasses all living humans and excludes all archaic humans (Lieberman et al. 2002), the generally agreed-upon physical characteristics of anatomical modernity include a high rounded skull, facial retraction, and a light and gracile, as opposed to heavy and robust, skeleton (Lahr 1996). Early fossils with these characteristics have been found in eastern Africa and have been dated to ~160,000–200,000 years ago (White et al. 2003; McDougall et al. 2005). At that time, the population of anatomically modern humans appears to have been small and localised (Harpending et al. 1998). Much larger populations of archaic humans lived elsewhere in the Old World, including the Neanderthals in Europe and an earlier species of humans, *Homo erectus*, in Asia (Swisher et al. 1994).
>
> Fossils of the earliest anatomically modern humans found outside Africa are from two sites in the Middle East and date to a period of relative global warmth, ~100,000 years ago, though this region was reinhabited by Neanderthals in later millennia as the climate in the

northern hemisphere again cooled (Lahr and Foley 1998). Groups of anatomically modern humans appear to have moved outside Africa permanently sometime >60,000 years ago. One of the earliest modern skeletons found outside Africa is from Australia and has been dated to ~42,000 years ago (Bowler et al. 2003), although studies of environmental changes in Australia argue for the presence of modern humans in Australia >55,000 years ago (Miller et al. 1999). To date, the earliest anatomically modern skeleton discovered from Europe comes from the Carpathian Mountains of Romania and is dated to 34,000–36,000 years ago (Trinkaus et al. 2003). (Berg et al., 'The use of racial, ethnic, and ancestral categories in human genetics research', pp. 519–32.)

89 **there is roughly ten times more genetic variation within large populations than there is between populations.**

⟶ Moreover, genetic variation is even higher inside Africa than it is elsewhere. The following data are according to researchers at the National Human Genome Research Institute:

> In general, however, 5–15 % of genetic variation occurs between large groups living on different continents, with the remaining majority of the variation occurring within such groups (Lewontin 1972; Jorde et al. 2000a; Hinds et al. 2005) . . .
>
> For example, ~90 % of the variation in human head shapes occurs within every human group, and ~10 % separates groups, with a greater variability of head shape among individuals with recent African ancestors (Relethford 2002).
>
> In addition to having higher levels of genetic diversity, populations in Africa tend to have lower amounts of linkage disequilibrium than do populations outside Africa. (Berg et al., 'The use of racial, ethnic, and ancestral cate-gories in human genetics research', pp. 519–32.)

It has also been determined that human beings are far less different from one another than other animals are within their own species:

> The data gathered to date suggest that human variation exhibits several distinctive characteristics. First, compared with many other mammalian species, humans are genetically less diverse [than other species]. (Berg et al., 'The use of racial, ethnic, and ancestral categories in human genetics research', pp. 519–32.)

89 **'While ancestry is a useful way to classify species':** Wilkins, 'Races, Geography, and Genetic Clusters'.

89 **By no stretch of the imagination, then, does any ethnicity or region have an exclusive lock on a particular body type or secret high-performance gene. Body shapes, muscle fibre types, etc., are actually quite varied and scattered, and true athletic potential is widespread and plentiful.**

—⊸ Even Jon Entine acknowledges this. Bob Young writes:

> Entine is careful to stress that he's talking about trends among groups of very elite athletes. He's not saying white guys should give up playing pickup ball because they can't jump. He is saying that among the small population of elite athletes, there are differences that could give a fraction-of-a-second advantage to people of African ancestry, which makes the difference, at the elite level, between a medal and fourth place . . .
>
> In the end, Entine says, the individual's work ethic, competitive spirit and training remain the key to success. 'That's why plenty of guys with Scottie Pippen's talent are [stuck] in the CBA [Continental Basketball Association],' he says. (Young, 'The Taboo of Blacks in Sports'.)

90 **In the words of King's College's developmental psychopathologist Michael Rutter, genes are 'probabilistic rather than deterministic':** Rutter, Moffitt, and Caspi, 'Gene-environment interplay and psychopathology', pp. 226–61.

—⊸ For my critique of the term 'probabilistic,' see the note 'Many scientists have understood this much more complicated truth for years but have had trouble explaining it to the general public. It is, after all, a lot harder to explain and understand than simple genetic determinism' on page 151.

90 **A seven- or fourteen- or twenty-eight-year-old outfitted with a certain height, shape, muscle-fibre proportion, and so on is not that way merely because of genetic instruction.**

—⊸ Some of the truly fascinating insights into talent and greatness emerge from the realm of human musculature – how our skeletal muscles are initially formed, the attributes of different muscle fibres, and the different ways muscles can be transformed by activity and training. Reviewing the nature and nurture of muscles is also perhaps the best window into the dynamics of genetic expression. Here's an overview:

> The human body contains three basic muscle types:

- Smooth (involuntary muscles serving the digestive system, blood vessels, airways, etc.)
- Cardiac (also involuntary; cardiac muscle is self-excitable and designed to function on its own)
- Skeletal (all voluntary muscles, from eyes to fingers to toes).

This overview concentrates on skeletal muscles – the muscles we exert direct control over. Each skeletal muscle is a bundle of thousands of specialised elongated cells called muscle fibres.

The fibres are fed by tiny blood-filled capillaries, held together with various kinds of connective tissue, and fired ('innervated') by motor neurons – one neuron firing six hundred or so muscle fibres.

Each individual muscle fibre also contains a string of DNA-filled nuclei positioned just underneath and along the entire length of its membrane. The genetic material constantly instructs each fibre how to react and adapt to various circumstances.

There are two basic types of muscle fibres:

- 'Slow-twitch' (type I) fibres are designed to contract for long periods of time; packed with mitochondria, they are extremely efficient at converting oxygen to fuel. These fibres enable us to jog, swim, bicycle, and engage in other lengthy activities.
- 'Fast-twitch' (type II) fibres contract rapidly and forcefully for a period of seconds, very quickly using huge amounts of (anaerobic) energy, becoming spent and needing to rest and replenish. These fibres enable us to sprint, jump, lift weights, and engage in other short-burst activities.

In musculature, we are not all created equal. Although on average, human beings have about a fifty-fifty mix of slow- and fast-twitch muscle fibres, some are born with differing proportions.

'The "average" healthy adult has roughly equal numbers of slow and fast fibres in, say, the quadriceps muscle in the thigh. But as a species, humans show great variation in this regard; we have encountered people with a slow fibre percentage as low as 19 per cent and as high as 95 per cent in the quadriceps muscle.' (Anderson et al., 'Muscle, Genes and Athletic Performance'.)

As anyone might logically expect from the above description of the fibre types, a higher proportion of one or another can offer certain potential advantages to highly trained athletes. Elite marathon runners and cyclists benefit from a higher proportion of slow-twitch fibres, for example, while sprinters benefit from a higher proportion of fast-twitch fibres. (Anderson et al., 'Muscle, Genes and Athletic Performance'.)

These genetic differences, however, must be put into careful context.

First, muscle fibre proportion is only one of many performance factors. On its own, it is not a good predictor of individual performance. (Quinn, 'Fast and Slow Twitch Muscle Fibres'.)

Second, muscles are tremendously adaptive to external stimulus, and are designed to be so. The muscles we are born with are merely default muscles – ready and waiting to be re-created in one or another particular direction by active use.

To understand how adaptation is literally built into our muscle DNA, let's look at all the things that happen as a result of training.

At any given time, each muscle is adapted to a status quo of activity and exertion – i.e., each muscle is exactly as big, strong, and efficient as it needs to be. When pushed just beyond the ordinary level of exertion, a number of physiological changes begin to unfold:

1. Neural response. The first measurable effect is an increase in the neural drive stimulating muscle contraction. Within just a few days, an untrained individual can achieve measurable strength gains resulting from 'learning' to use the muscle.

2. Genetic response makes muscle fibres more efficient. In response to extended (aerobic) exercise – e.g. jogging – there is a genetic response in the nucleus of each cell fibre that makes it more efficient and enduring, increasing the number of mitochondria and provoking an increase in surrounding capillaries and the accumulation of fats and carbohydrates.

3. Genetic response makes muscle fibres become stronger and grow in size. In response to overload/resistance exercise – e.g. weight lifting – the DNA responds with instructions that will lead to the strengthening and enlarging [hypertrophy] of each fibre.

> As the muscle continues to receive increased demands . . . upregulation appears to begin with the ubiquitous second messenger system (including phospholipases, protein kinase C, tyrosine kinase, and others). These, in turn, activate the family of immediate-early genes, including *c-fos, c-jun* and *myc*. These genes appear to dictate the contractile protein gene response.
>
> Finally, the message filters down to alter the pattern of protein expression. It can take as long as two months for actual hypertrophy to begin. The additional contractile proteins appear to be incorporated into

existing myofibrils (the chains of sarcomeres within a muscle cell) . . . These events appear to occur within each muscle fibre. That is, hypertrophy results primarily from the growth of each muscle cell, rather than an increase in the number of cells. (National Skeletal Muscle Research Center, 'Hypertrophy'.)

4. When training is particularly intense and prolonged, slow-twitch muscle fibres can become transformed into fast-twitch fibres, and vice versa.

> Adult skeletal muscle shows plasticity and can undergo conversion between different fibre types in response to exercise training or modulation of motoneuron activity. (Wang et al., 'Regulation of muscle fiber type and running endurance by PPAR'.)

A detailed diagram of gene expression at work in muscle fibres:

Exercise, stretches and other muscle activity (LEFT) interacts with DNA in the nucleus (CENTRE), which in turns interacts with protein translators to effect changes on the cell and surrounding tissue (RIGHT).

(Source of graphic and detailed explanation of genetic transcription: Rennie et al., 'Control on the size of the human muscle mass', p. 802.)

In sum, while evolution has given humans some variability in muscle types, perhaps its powerful product is its adaptivity. *Muscles are designed to be rebuilt*. 'The ability of striated muscle tissue to adapt to changes in activity or in working conditions is extremely high. In some ways it is comparable to the ability of the brain to learn.' (Bottinelli and Reggiani, eds., *Skeletal Muscle Plasticity in Health and Disease*.)

Citations

AMONG HUMANS, GREAT VARIATION IN MUSCLE-FIBER RATIOS

Anderson, Jesper L., Peter Schjerling, and Bengt Saltin. 'Muscle, Genes and Athletic Performance'. *Scientific American*, September 2000.

DIFFERENT FIBER RATIOS PROVIDE ADVANTAGES AND DISADVANTAGES FOR CERTAIN SPORTS

Anderson, Jesper L., Peter Schjerling, and Bengt Saltin. 'Muscle, Genes and Athletic Performance'. *Scientific American*, September 2000.

MUSCLE-FIBER TYPE IS A POOR PREDICTOR OF PERFORMANCE

Quinn, Elizabeth. 'Fast and Slow Twitch Muscle Fibers: Does Muscle Type Determine Sports Ability?' Published on the About.com Sports Medicine Web site, October 30, 2007.

Articles cited by Quinn for further reference

Anderson, Jesper L., Peter Schjerling, and Bengt Saltin. 'Muscle, Genes and Athletic Performance'. *Scientific American*, September 2000.

McArdle, W. D., F. I. Katch, and V. L. Katch. *Exercise Physiology: Energy, Nutrition and Human Performance.* Williams & Wilkins, 1996.

Lieber, R. L. *Skeletal Muscle Structure and Function: Implications for Rehabilitation and Sports Medicine.* Williams & Wilkins, 1992.

Thayer, R., J. Collins, E. G. Noble, and A. W. Taylor. 'A decade of aerobic endurance training: histological evidence for fibre type transformation'. *Journal of Sports Medicine and Physical Fitness* 40, no. 4 (2000): 284–89.

NEURAL RESPONSE AND GENETIC RESPONSE

National Skeletal Muscle Research Center. 'Hypertrophy'. Published on the UCSD Muscle Physiology Laboratory Web site.

GENETIC RESPONSE MAKES MUSCLE FIBERS MORE EFFICIENT

Russell, B., D. Motlagh, and W. W. Ashley. 'Form follows function: how muscle shape is regulated by work'. *Journal of Applied Physiology* 88, no. 3 (2000): 1127–32.

CONVERSION BETWEEN DIFFERENT FIBER TYPES

Wang, Yong-Xu, et al. 'Regulation of muscle fiber type and running endurance by PPAR'. Published on the Public Library of Science Web site, August 24, 2004.

Kohn, Tertius A., Birgitta Essén-Gustavsson, and Kathryn H. Myburgh. 'Do skeletal muscle phenotypic characteristics of Xhosa and Caucasian endurance runners differ when matched for training and racing distances?' *Journal of Applied Physiology* 103 (2007): 932–40.

Coetzer, P., T. D. Noakes, B. Sanders, M. I. Lambert, A. N. Bosch, T. Wiggins, and S. C. Dennis. 'Superior fatigue resistance of elite black South African distance runners'. *Journal of Applied Physiology* 75 (1993): 1822–27.

Andersen, J. L., H. Klitgaard, and B. Saltin. 'Myosin heavy chain isoforms in single fibres from m. vastus lateralis of sprinters: influence of training'. *Acta Physiologica Scandinavica* 151 (1994): 135–42.

Pette, D., and G. Vrbova. 'Adaptation of mammalian skeletal muscle fibers to chronic electrical stimulation'. *Reviews of Physiology, Biochemistry and Pharmacology* 120 (1992): 115–202.

Trappe, S., M. Harber, A. Creer, P. Gallagher, D. Slivka, K. Minchev, and D. Whitsett. 'Single muscle fiber adaptations with marathon training'. *Journal of Applied Physiology* 101 (2006): 721–27.

90–91 This nongenetic aspect of inheritance is often overlooked by genetic determinists: culture, knowledge, attitudes, and environments are also passed on in many different ways: See chapter 7.

91 ***'The large variance in both the global and individual admixture estimates'*:** Benn-Torres et al., 'Admixture and population stratification in African Caribbean populations', pp. 90–98.

91 **The annual high school Boys' and Girls' Athletic Championships:** Rastogi, 'Jamaican Me Speedy'.

91 **'dozens of small children showed up for a Saturday morning youth track practice':** Layden and Epstein, 'Why the Jamaicans Are Running Away with Sprint Golds in Beijing'.

92 **Dennis Johnson did come back to Jamaica to create a college athletic programme:** Clark, 'How Tiny Jamaica Develops So Many Champion Sprinters'; Rastogi, 'Jamaican Me Speedy'.

92 **'We genuinely believe that we'll conquer,' says Jamaican coach Fitz Coleman:** Clark, 'How Tiny Jamaica Develops So Many Champion Sprinters'.

92 **a person's mind-set has the power to dramatically affect both short-term capabilities and the long-term dynamic of achievement:** Dweck, *Mindset*; Elliot and Dweck, eds., *Handbook of Competence and Motivation.*

92–93 Bannister himself later remarked that while biology sets ultimate limits to performance, it is the mind that plainly determines how close individuals come to those absolute limits.

'Though physiology may indicate respiratory and cardiovascular limits to muscular effort,' commented Bannister, 'psychological and other factors beyond the ken of physiology set the razor's edge of defeat or victory and determine how closely the athlete approaches the absolute limits of performance.' (Bannister, 'Muscular effort', pp. 222–25.)

There's also a national pride that works both to give Kenyan runners a psychological boost and to intimidate non-Kenyans. The emergent aura of invincibility around the Kenyan runners 'cannot be overestimated,' says sports psychologist Bruce Hamilton. (Hamilton, 'East African running dominance', p. 393.)

93 **'The past century has witnessed a progressive, indeed remorseless improvement in human athletic performance':** Noakes, 'Improving Athletic Performance or Promoting Health Through Physical Activity'.

Actual record times for the mile: 4:36.5 (1865), 3:43.13 (1999). Infoplease.com.

93 **The one-hour cycling distance record increased from 26 kilometres in 1876 to 49 kilometres in 2005.**

March 25, 1876, F. L. Dodds, 26.5 kilometres (Burke, *High-tech Cycling*.)

July 19, 2005, Ondrej Sosenka, 49.7 kilometres (Willoughby, 'Czech Ondrej Sosenka Sets New World One-hour Cycling Record of 49.7 km'.)

93 **The 200-metre freestyle swimming record decreased from 2:31 in 1908 to 1:43 in 2007.**

Actual times: 2:31.6, 1:43.86. (Agenda Diana swimming records Web site.)

93 **Technology and aerodynamics are a part of the story, but the rest of it has to do with training intensity, training methods, and sheer competitiveness and desire.**

University of Cape Town sports biologist Timothy David Noakes lists his '15 Laws of Training':

1. Train frequently all year round.
2. Start gradually and train gently.
3. Train first for distance, only later for speed.
4. Don't set yourself a daily schedule.
5. Alternate hard and easy training.
6. At first, try to achieve as much as possible on a minimum of training.
7. Don't race in training, and run time-trials and races longer than 16 km only infrequently.
8. Specialise.

9. Incorporate base training and peaking (sharpening).
10. Don't overtrain.
11. Train under a coach.
12. Train the mind.
13. Rest before a big race.
14. Keep a detailed logbook.
15. Understand the holism of training.

Noakes, 'Improving Athletic Performance or Promoting Health Through Physical Activity'.

93 **They are participants in a culture of the extreme, willing to devote more, to ache more, and to risk more in order to do better.**
—◦⪢ In the late twentieth and early twenty-first centuries, the extreme athletic culture has yielded both short-term dangers (such as 'overtraining syndrome') and long-term dangers such as premature skeletal aging and psychological damage. (Budgett, 'ABC of sports medicine', 465–68.)

CHAPTER 7:
HOW TO BE A GENIUS (OR MERELY GREAT)

PRIMARY SOURCES

Oyama, Susan, Paul E. Griffiths, and Russell D. Gray. *Cycles of Contingency: Developmental Systems and Evolution*. MIT Press, 2003.

Csikszentmihályi, Mihály, Kevin Rathunde, and Samuel Whalen. *Talented Teenagers*. Cambridge University Press, 1993.

CHAPTER NOTES

97 **'Are [people] conceived with the capacity to play a number of qualitatively different developmental tunes':** Bateson, 'Behavioral Development and Darwinian Evolution', p. 153.

98 **'SKYLAR: How did you do that?'** *Good Will Hunting*. Directed by Gus Van Sant. Big Gentlemen Limited Partnership. 1998.

98 **Neighbours of the Beethovens . . . recall seeing a small boy:** Morris, *Beethoven*, p. 16.

99 **Today, talk of giftedness still pervades our language, even among scientists who know better.**

David Moore writes:

> It appears that merely comprehending what genes actually do does not necessarily lead to a rejection of genetic determinism, because in spite of evidence to the contrary, even some biologists continue to write as if developmental processes can be genetically determined. (Moore, 'Espousing interactions and fielding reactions', p. 332.)

100 Even in a land of free choice, we are mostly shaped by habits, messages, schedules, expectations, social infrastructure, and natural surroundings that are not exclusively our own. Many of these elements are passed down from generation to generation with little or no change and are difficult or impossible to alter.

—◦⪢ Many people who stand out as being extraordinary do so because of choices they have made to stand radically apart from cultural norms: they may allocate time and resources in a very different way from their friends and neighbours.

101 'talent is much more widely distributed than its manifestation would suggest': Csikszentmihályi, Rathunde, and Whalen, *Talented Teenagers*, p. 2.

102 The source of motivation is often mysterious, but not always. One of the quirks of human emotion and psychology is that deep motivation can have more than one possible origin. A person can become joyfully inspired, spiritually devoted, or deeply resentful; motivation can be selfish or vengeful, or arise out of a desperation to prove someone right or wrong; it can be conscious or unconscious.

Mihály Csikszentmihályi suggests two very different points of origin:

> The relationship between early family environment and later creative achievement is rather ambiguous. On the one hand, a context of optimal support and stimulation seems necessary. On the other hand, the lives of some of the greatest creative geniuses contradict this notion, being full of early trauma and tragedy. On the basis of longitudinal studies of young artists and talented adolescents, as well as a retrospective study of mature creative individuals, we explore the outcomes of various family environments. It seems that the two extremes of optimal and pathological experience are both represented disproportionately in the backgrounds of creative individuals. However, creative persons whose childhood was more traumatic appear less satisfied with themselves and

their work. So, although a difficult childhood might be more conducive to creative achievement, it does not seem to lead to a serene adulthood. Our study of talented teenagers showed that students who came from a 'complex' family environment that provided them with both support and stimulation were more likely to take on new challenges in their area of talent and to enjoy working on and developing their skills. Such students reported feeling happy more often than those from other family types, and were significantly happier when spending time alone or in productive work. (Csikszentmihályi and Csikszentmihályi, 'Family influences on the development of giftedness', pp. 187–200.)

102 They wish they had done more: got more education, worked harder, persevered: Hattiangadi, Medvec, and Gilovich, 'Failing to act', pp. 175–85.

103–104 'I wake up sometimes and say, "What the heck happened to me?" It's like a nightmare,' American runner Abel Kiviat told the *Los Angeles Times* in 1990 about his disappointing silver medal in the 1,500-metre Olympic run. Kiviat was ninety-one when he made this statement – his performance had occurred more than seventy years earlier: Medvec, Madey, and Gilovich, 'When less is more', p. 609.

106 Charles Darwin had so little to show for himself as a teenager that his father said to him, 'You care for nothing but shooting, dogs, and rat-catching and you will be a disgrace to yourself and all your family.'
—⊸ At age twenty-two, Charles Darwin set out on the HMS *Beagle*, embarking on a voyage that would lead to one of the most important scientific theories in human history. (Simonton, *Origins of Genius*, p. 109.)

106 To know the particulars of a favourite artist or athlete's ordeal is to be continually reminded of uncharted paths and oddball ideas that only later become recognised as genius. This experience is magnified by examining rough drafts of masterpiece books, paintings, and albums.
—⊸ Prime examples of a great work of art in progress:

- Peter F. Neumeyer's *The Annotated Charlotte's Web* – a line-by-line look at all the work that went into E. B. White's *Charlotte's Web*.
- The Beatles' legendary 'Strawberry Fields' demos.

106 'how beautiful things grow out of shit': Brian Eno, in the Daniel Lanois documentary *Here Is What It Is*.

107 'Most students who become interested in an academic subject do so because they have met a teacher who was able to pique their interest': Csikszentmihályi, Rathunde, and Whalen, *Talented Teenagers*, p. 7.

→ As for me, I've been lucky enough to have several life-changing teachers:

Mrs. Beti Gould, preschool and kindergarten
Mr. Giovanni Mucci, third grade
Mr. Bob Moses, eighth grade and eleventh grade
Mrs. Marie King Johnson, eleventh and twelfth grade
Professor Andrew Hoffman, first year of college

CHAPTER 8:
HOW TO RUIN (OR INSPIRE) A KID

PRIMARY SOURCES

Csikszentmihályi, Mihály, Kevin Rathunde, and Samuel Whalen. *Talented Teenagers*. Cambridge University Press, 1993.

Gardner, Howard. 'Do Parents Count?' *New York Review of Books*, November 5, 1998.

Harper, Lawrence V. 'Epigenetic inheritance and the intergenerational transfer of experience'. *Psychological Bulletin* 131, no. 3 (2005): 340–60.

Harris, Judith Rich. *The Nurture Assumption: Why Children Turn Out the Way They Do*. Simon & Schuster, 1999.

Turkheimer, Eric. 'Three laws of behavior genetics and what they mean'. *Current Directions in Psychological Science* 9, no. 5 (October 2000): 160–64.

CHAPTER NOTES

108 Do we know how many geniuses are never recognised: Csikszentmihályi, Rathunde, and Whalen, *Talented Teenagers*, p. 2.

108 In 1999, Oregon neuroscientist John C. Crabbe led a study: Crabbe, Wahlsten, and Dudek, 'Genetics of mouse behavior', pp. 1670–72.

109 This was unforeseen, and it turned heads.

Google Scholar shows 556 scientific articles/books referencing this one article.

110 What we do know is that our brains and bodies are primed for plasticity.

In *Resiliency*, Bonnie Benard writes, 'Findings over this past decade [point] to the plasticity of the human brain' (Bruer, 1999; Diamond & Hopson, 1998; Ericsson et al., 1998; Kagan, 1998). As Daniel Goleman notes in his discussion of the 'protean brain', the 'finding that the brain and nervous system generate new cells as learning or repeated experiences dictate has put the theme of plasticity [*emphasis added*] at the front and center of neuroscience' (2003, p. 334). Unfortunately, what the public has been left with instead, warns prominent developmental psychologist Jerome Kagan, is the 'seductive' notion of 'infant determinism' (1998).

Benard's Citations

Benard, Bonnie. *Resiliency: What We Have Learned.* WestEd, 2004.

Bruer, J. *The Myth of the First Three Years*. Free Press, 1999.

Diamond, M., and J. L. Hopson. *Magic Trees of the Mind: How to Nurture Your Child's Intelligence, Creativity, and Healthy Emotions from Birth Through Adolescence.* Penguin, 1999.

Kagan, J. *Three Seductive Ideas*. Harvard University Press, 1998.

Goleman, Daniel. *Destructive Emotions: A Scientific Dialogue with the Dalai Lama*. Bantam, 2003.

110 'Recent reviews of pre- and postnatal brain': Johnson and Karmiloff-Smith, 'Neuroscience Perspectives on Infant Development', p. 123. This entire chapter is highly recommended, and may be accessed online via Google Books. Go to 'Contents', and click on page 121.

110 'Human babies are special': Meltzoff, 'Theories of People and Things'.

111 Musical ability lies dormant in all of us, calling for early and sustained incantation.

See earlier note 'Levitin also concurs with University of California, San Diego's Diana Deutsch' on page 235.

111 Based on our reading of these interactions, we then tailor her environment accordingly.

In his landmark book *Touchpoints*, paediatrician T. Berry Brazelton writes:

> There are wide individual differences in the style in which a baby handles responses to stimuli around him, in his need for sleep and his crying. Babies differ in how they can be soothed, as well as in their responses to hunger and discomfort, to exposure to temperature changes, to handling, and to interaction with caregivers. The task for parents . . . [is] to watch and listen for their own baby's particular style. (Brazelton, *Touchpoints*, 1992.)

112 Challenging assumptions is always healthy, and in one sense Harris's book was a welcome critique that forced university psychologists out of their comfort zone.

Howard Gardner writes:

> As Harris shrewdly points out, there are two problems with the nurture assumption. First, when viewed with a critical eye, the empirical evidence about parental influences on their children is weak, and often equivocal. After hundreds of studies, many with individually suggestive findings, it is still difficult to pinpoint the strong effects that parents have on their children. Even the effects of the most extreme experiences – divorce, adoption, and abuse – prove elusive to capture. Harris cites Eleanor Maccoby, one of the leading researchers in the field, who concluded that 'in a study of nearly four hundred families, few connections were found between parental child-rearing practices (as reported by parents in detailed interviews) and independent assessments of children's personality characteristics – so few, indeed, that virtually nothing was published relating the two sets of data.' (Gardner, 'Do Parents Count?')

112 'Genes contain the instructions for producing a physical body and a physical brain': Harris, *The Nurture Assumption*, p. 30.

112 'non-shared' environment – a term proposed by geneticist Robert Plomin to explain not-yet-understood environmental influences.

Catherine Baker writes:

> The well-known geneticist Robert Plomin and a colleague first posed this question in an article published in 1987 (R. Plomin and D. Daniels 1987, *Behavioral and Brain Sciences* 10: 1–60). They proposed this answer: the differences result from aspects of the environment that siblings raised together do not share. They termed this the non-shared environment. So, for example, socioeconomic status such as poverty would be a shared environmental influence while illness, specific

traumatic events, or parental attitudes towards each individual child would be non-shared environmental influences. The concept of a non-shared environment launched a wave of studies seeking to identify the variables within a family environment that differ for each sibling. (Baker, Report on Eric Turkheimer's presentation 'Three laws of behavior genetics and what they mean'; Baker references Plomin and Daniels, 'Why are children in the same family so different from one another?' pp. 1–60.)

113 Two years after her book came out, though, it turned out that there was a problem with the shared/non-shared paradigm. An analysis in 2000 by the University of Virginia's Eric Turkheimer revealed that it was another false distinction. Just like 'nature/nurture' was supposed to separate genetic effects from environmental effects, 'shared' and 'non-shared' implied that it was either/or: either people would have similar reactions to shared experiences *or* they would have different reactions to non-shared experiences. Turkheimer's powerful meta-analysis revealed the much more common third possibility: most of the time, kids have different reactions to shared experiences.

From Turkheimer's paper:

> Again and again, Plomin and his colleagues have emphasised that the importance of nonshared environment implies that it is time to abandon shared environmental variables as possible explanations of developmental outcomes. And although modern environmentalists might not miss coarse measures like socioeconomic status, it is quite another thing to give up on the causal efficaciousness of normal families, as Scarr (1992), Rowe (1994), and Harris (1998) have urged. If, however, nonshared environmental variability in outcome is the result of the unsystematic consequences of both shared and nonshared environmental events, the field faces formidable methodological problems – Plomin and Daniels's gloomy prospect – but need not conclude that aspects of families children share with siblings are of no causal importance. (Turkheimer, 'Three laws of behavior genetics and what they mean'.)

113 Harvard psychologist Howard Gardner had an even more fundamental problem with Harris's notion of uninfluential parents. 'When we consider the empirical part of Harris's argument,' he wrote in the *New York Review of Books*, 'we find it is indeed true that the research on parent-child socialisation is not what we would hope for. However, this says less about parents and children and more about the

state of psychological research, particularly with reference to "softer variables" such as affection and ambition. While psychologists have made genuine progress in the study of visual perception and measurable progress in the study of cognition, we do not really know what to look for or how to measure human personality traits, individual emotions, and motivations, let alone character.'

Gardner continues:

> Consider, as an example, the categories that the respondents must use when they describe themselves or others on the Personal Attributes Questionnaire . . . [They] are asked whether they would describe themselves as Gentle, Helpful, Active, Competitive, and Worldly. These terms are not easy to define and people are certainly prone to apply them favourably to their own case. Or consider the list of acts from which observers can choose to characterise children from different cultures – Offers Help, Acts Sociably, Assaults Sociably, Seeks Dominance . . . We don't know with any confidence what these acts mean to children, adolescents, and adults in diverse cultures. (Gardner, 'Do Parents Count?')

114 'I would give much weight to the hundreds of studies pointing toward parental influence and to the folk wisdom accumulated by hundreds of societies over thousands of years.'

At this point in his paper, Gardner adds his own footnote:

> Make that one more. To be published in February 1999 is Frank Furstenberg et al., *Managing to Make It: Urban Families and Adolescent Success* (University of Chicago Press). Directly countering a claim by Harris, this sociological study indicates that neighbourhood has surprisingly little effect on the relative success or failure achieved by adolescents. Rather, consistent with common sense and many other psychological and sociological studies, the research team finds that the parents of successful adolescents 'continued to be active agents looking out for their children's interests throughout adolescence.' They knew which resources were available and used them, they encouraged some interests and discouraged others, they organised family activities, spent informal time with the youngsters, and knew enough to cut the youths a certain amount of slack. (Gardner, 'Do Parents Count?')

114 So yes, parents matter. Parenting isn't everything or the only thing. Parents don't have anything close to complete control and in most

cases should not shoulder all the blame when things don't turn out well. But parenting does matter.

Lawrence Harper points out a favourite study supporting this argument:

> On the one hand, the evidence clearly shows that parenting influences do matter. For example, Sroufe (2002) reported striking results from a long-term longitudinal study of low-socioeconomic-status families. He found that early quality of care predicted a range of later outcomes including competence in peer relations, adolescent risk taking, emotional problems, and school success. In the latter case, a composite of six measures of quality of parenting, the home environment, and the quality of stimulation afforded the child could predict high school dropout with 77% accuracy. (Harper, 'Epigenetic inheritance and the intergenerational transfer of experience', pp. 340–60.)

114 'Isn't that something of an accomplishment?': Suzuki, *Nurtured by Love*, p. 1.

115 He came to quickly believe, in fact, that early musical training has an overwhelming advantage over later training, and that it was a gateway to an enlightened life.

Suzuki friend and biographer Evelyn Hermann gives us the following quote from him: 'I am not very interested in doing "repair" work on people who can play already,' he wrote to a fellow instructor in 1945. 'What I want to try is infant education.' (Hermann, *Shinichi Suzuki*, p. 38.)

115 'talent is not inherent or inborn, but trained and educated': Hermann, *Shinichi Suzuki*, p. 40.

115 his Talent Education Research Institute had thirty-five branches in Japan and was teaching fifteen hundred children: 'Personal History of Shinichi Suzuki'.

115 The Suzuki method became a sensation around the world and helped transform our understanding of young children's capabilities.

In his 1969 autobiography, Suzuki relayed the story of Peeko the coughing parakeet:

> Peeko lived in the Tokyo classroom of Suzuki instructor Mr. Miyazawa, who had painstakingly taught his bird to say: 'I am Peeko Miyazawa,' and 'Peeko is a good little bird.' Achieving these results was

> simply a matter of repetition and persistence, much the same as with children and violins.
>
> But then it got really interesting. One week, Mr. Miyazawa had a bad cold for several days and coughed a lot in class. Lo and behold, Peeko the parakeet began to follow 'I am Peeko Miyazawa' with a distinct cough sound. He also began to hum 'Twinkle Twinkle Little Star', after hearing the students play it over and over again on the violin.
>
> 'Talent develops talent,' concluded Suzuki. 'The planted seed of ability grows with ever increasing speed.' (Suzuki, *Nurtured by Love*, p. 6.)

This is the positive feedback loop that I referred to in chapter 6 – what Lawrence Harper calls a 'self-amplifying cycle'.

116 In early adulthood, Freed explains, the child will inevitably struggle with social and emotional challenges (as everyone does) and find that he doesn't have a deep emotional reservoir to fall back on. The foundations of love and trust are corrupted by what he experienced as a child. The child victim of a narcissistic parent frequently has a difficult time forming stable life partnerships: Conversation with Dr. Peter Freed, November 8, 2008.

Joan Freeman also mentions a study that seems to touch on this same syndrome:

> A 15-year Chinese study of 115 extremely high-IQ children showed the strong influence of family provision, both in achievement and emotional development. The children were first identified by parents then validated as gifted by a psychologist. Every year the parents were interviewed several times. By the age of three many children could recognise 2000 Chinese characters, and at four many could not only read well, but also wrote compositions and poems. However, these 'hothoused' children were found to lack easy social relationships so the parents were given lessons in how to help their children get on with others. (Freeman, 'Giftedness in the long term', 384–403.)

117 a parent must not use affection as a reward for success or a punishment for failure.

As seen in a recent study by Echo H. Wu. (Wu, 'Parental influence on children's talent development', pp. 100–129.)

117 Persistence, she argues, 'must have an inborn, biological component': Von Károlyi and Winner, 'Extreme Giftedness', p. 379.

117 The brain circuits that modulate a person's level of persistence are plastic – they *can* be altered.
—⊕⊗ Looking at MRI scans, researchers have even been able to see varying levels of persistence light up in specific regions of the brain. (Gusnard et al., 'Persistence and brain circuitry', pp. 3479–84.)

The Robert Cloninger comment was made to Po Bronson. Cloninger, at Washington University in St. Louis, not only zeroed in on the persistence circuitry in the brain but also trained mice and rats to develop persistence. According to Cloninger, the animal mind can actually be trained to reward itself for slow and steady progress rather than the more thrilling instant gratification. (Bronson, 'How Not to Talk to Your Kids'.)

118 a classic study by Stanford psychologist Walter Mischel.
More from the marshmallow experiments:

> Observation of children's spontaneous behaviour during the delay process also suggested that those who were most effective in sustaining delay seemed to avoid looking at the rewards deliberately, for example, covering their eyes with their hands and resting their heads on their arms. Many children generated their own diversions: they talked quietly to themselves, sang, created games with their hands and feet, and even tried to go to sleep during the waiting time. Their attempts to delay gratification seemed to be facilitated by external conditions or by self-directed efforts to reduce their frustration during the delay period by selectively directing their attention and thoughts away from the rewards. However, it also seemed unlikely that sheer suppression or distraction from the frustration caused by the situation is the only determinant of this type of self-control. Indeed, when certain types of thoughts are focused on the rewards they can facilitate self-control substantially, even more than distraction does, as the next set of experiments found.
>
> The results so far show that exposure to the actual rewards or cues to think about them undermine delay, but the studies did not consider directly the possible effects of images or symbolic representations of rewards. Yet it may be these latter types of representation – the images of the outcomes, rather than the rewards themselves – that mediate the young child's ability to sustain delay of gratification. To explore this possibility, the effects of exposure to realistic images of the rewards were examined by replicating the experiments on the effects of reward exposure with slide-presented images of the rewards. It was found that

although exposure to the actual rewards during the delay period makes waiting difficult for young children, exposure to images of the rewards had the opposite effect, making it easier. Children who saw images of the rewards they were waiting for (shown life-size on slides) delayed twice as long as those who viewed slides of comparable control objects that were not the rewards for which they were waiting, or who saw blank slides. Thus, different modes of presenting rewards (that is, real versus symbolic) may either hinder or enhance self-control. (Mischel, Shoda, and Rodriguez, 'Delay of gratification in children', p. 935.)

119 Don't immediately respond to their every plea. Let them learn to deal with frustration and want. Let them learn how to soothe themselves and discover that things will be all right if they wait for what they want.

→ An excellent article on this subject:

Who's in Charge? Teach Kids Self-Control
By Jennifer Keirn

In the early years, it's easy to know who's in control. Parents sit front and centre in the cockpit, steering children through infancy and toddlerhood. Parents control where kids go, what they do and with whom, what they eat and what they wear. That's not to say they don't throw some turbulence our way – some more frequently than others – but it's our job as parents to steady the aircraft and return to course.

As our kids grow, however, we're faced with the challenge of gradually loosening our commanding grip on the controls. In the end, it's our kids, not us, who will land their plane, and they've got to know how to control it for themselves, make the right decisions and resist negative impulses.

Teaching kids self-control is one of the most important things we can do to prepare them for life, yet it's also one of the most difficult. Dr. Sylvia Rimm, child psychologist and director of the Family Achievement Clinic in Westlake, says teaching this critical life skill requires putting a combination of good parenting principles to work.

Be a good role model. 'They're watching you all the time,' Rimm says. 'Self-control is a lot of things. Do you buy what you want when you want it regardless of price? Do you eat or drink whatever you want regardless of the consequences?' That's why it's critical for parents to model self-control if they want their kids to learn it, she adds.

Give kids practice in delayed gratification. Research has shown that the ability to delay gratification in childhood is an indicator of success later in life . . . 'Self-control is built through delayed gratification,'

Rimm says. 'That means early on, if you give in when your toddler or preschooler cries because they want something, you're not teaching them self-control.'

Practise consistent parenting. 'The whole parenting team has to be united with each other in setting limits,' she says. 'If one parent says "yes" and the other parent says "no", kids won't learn self-control. They'll just learn how to manipulate their parents.'

Set age-appropriate limits. Rimm encourages parents to visualise the 'v' in 'love' as a tool for setting limits as a child grows. Children begin at the bottom of the 'v' in infancy and toddlerhood, with limited freedom and choices. As they grow, moving up the 'v', parents should gradually allow more freedom and power while still providing parental limits.

Teaching kids self-control isn't like teaching them to tie their shoes or use the potty. Rimm says teaching self-control is a gradual process that should begin in infancy and continue through the teenage years. Each lesson builds on the previous one, making it critical for parents to lay the foundations for self-control early on.

'A lack of self-control, of not learning delayed gratification, is tied directly into kids getting involved in alcohol, sex and drugs in the teen years,' Rimm says.

Rimm offers tips for teaching self-control at each developmental stage:

Toddlers and preschoolers. 'Toddlers through school-age kids are very concrete,' she says. 'Things are very black-and-white.' Setting limits at this stage should consist of 'yes' or 'no' responses, no in-betweens.

Children also directly imitate their parents at this stage, making it important for parents to start modelling self-control from the start. Parents also can begin teaching kids to delay gratification by not giving in when they cry for something.

School-age kids. As the 'v' gradually widens and school-age kids begin to have more choices and freedoms, they get to put those developing self-control skills to work in daily life. 'When kids have chores and allowance, they can start saving their money and start counting the days until Christmas or their birthday,' Rimm says. 'This is how they learn delayed gratification.' This is the stage at which kids will begin to nag you for what they want, as opposed to crying for it in their earlier years.

Pre-teens. 'In middle school, kids are now exposed to environments more like their parents encountered in high school and college,' she says. Drugs, sex and alcohol are finding their way into younger ages than ever, testing children's developing self-control skills. 'Kids who haven't learned positive influences earlier are more easily pulled into the drug scene. Parents need to help kids search out appropriate peers and do a lot of fun, family activities to balance out what they're getting at school. You can't just be a family that says "no" all the time; you need to be a family that has fun.'

Teenagers. In the teen years, kids should be approaching the top of the 'v', getting ready to move on to adult independence and personal decision-making. This also is the stage at which peer pressure reaches its peak, as do negative influences that require good self-control.

'Right through the teen years, parents have to be sure to set limits,' Rimm says. 'Hormones are rushing, and they're surrounded by movies, TV, peers. It was much easier to be self-controlled a few generations ago.'

If parents haven't been building the foundations of self-control, she says, the teen years are the toughest time to start, and parents may need help. 'If parents understand what's gone wrong, they can fix it, but more extreme cases should go to family therapy. Parents who want to say "no" will get support from the therapist.'

Perhaps you're the parent of a newborn and in complete control of the cockpit, or perhaps you're the parent of a just-graduated senior who's ready to take over the controls and fly off to college. Regardless, self-control is a life skill to be taught and reinforced again and again, to ensure your children make a safe landing into adulthood. (Keirn, 'Who's in Charge? Teach Kids Self-Control'.)

119 'Specific motor problems': Reed and Bril, 'The Primacy of Action in Development', p. 438.

CHAPTER 9:
HOW TO FOSTER A CULTURE OF EXCELLENCE

PRIMARY SOURCES

Durik, Amanda M., and Judith M. Harackiewicz. 'Achievement goals and intrinsic motivation: coherence, concordance, and achievement orientation'. *Journal of Experimental Social Psychology* 39, no. 4 (2003): 378–85.

Gneezy, Uri, Kenneth L. Leonard, and John A. List. 'Gender Differences in Competition: The Role of Socialization'. UCSB Seminar Paper. Published on the University of California, Santa Barbara, Department of Economics Web site. June 19, 2006.

Goffen, Rona. *Renaissance Rivals: Michelangelo, Leonardo, Raphael, Titian.* Yale University Press, 2004.

Mighton, John. *The Myth of Ability: Nurturing Mathematical Talent in Every Child.* Walker, 2004.

Tauer, John M., and Judith M. Harackiewicz. 'Winning isn't everything: competition, achievement orientation, and intrinsic motivation'. *Journal of Experimental Social Psychology* 35 (1999): 209–38.

CHAPTER NOTES

122 da Vinci sported a public 'disdain' for his younger peer Michelangelo Buonarroti – a hostility so strong that the great Michelangelo eventually felt compelled to leave Florence so that he and Leonardo wouldn't have to share the same town.

According to Giorgio Vasari:

> There was very great disdain between Michelangelo Buonarroti and him, on account of which Michelangelo departed from Florence, with the excuse of Duke Giuliano, having been summoned by the Pope to the competition for the façade of S. Lorenzo. Leonardo, understanding this, departed and went into France, where the King, having had works by his hand, bore him great affection; and he desired that he should colour the cartoon of S. Anne, but Leonardo, according to his custom, put him off for a long time with words.
>
> Finally, having grown old, he remained ill many months, and, feeling himself near to death, asked to have himself diligently informed of the teaching of the Catholic faith, and of the good way and holy Christian religion; and then, with many moans, he confessed and was penitent; and although he could not raise himself well on his feet, supporting himself on the arms of his friends and servants, he was pleased to take devoutly the most holy Sacrament, out of his bed. The King, who was wont often and lovingly to visit him, then came into the room; wherefore he, out of reverence, having raised himself to sit upon the bed, giving him an account of his sickness and the circumstances of it, showed withal how much he had offended God and mankind in not having worked at his art as he should have done. Thereupon he was seized by a paroxysm, the messenger of death; for which reason the King having risen and having taken his head, in order to assist him and show him favour, to the end that he might alleviate his pain, his spirit, which was divine, knowing that it could not have any greater honour, expired in the arms of the King, in the seventy-fifth year of his age. (Vasari, 'Life of Leonardo da Vinci', pp. 104–5.)

122 Da Vinci also pointedly criticised the art of sculpture – Michelangelo's forte – as a messy, easier, and obviously inferior craft that requires 'greater physical effort [while] the painter conducts his works with greater mental effort.'

All of this is from Leonardo's own *Paragone (A comparison of the Arts)*. (Goffen, *Renaissance Rivals*, p. 65.)

For specifics about 'messy' and 'easier', see 'Paragone: painting or sculpture?' on the Universal Leonardo Web site.

122 Walking with a friend near S. Trinità: Symonds, *The Life of Michelangelo Buonarroti*, p. 173.

123 'Every natural gift must develop itself by contests,' wrote Nietzsche.

More: 'Without envy, jealousy, and competing ambition the Hellenic State like the Hellenic man degenerates. He becomes bad and cruel, thirsting for revenge, and godless; in short, he becomes "pre-Homeric".' (Nietzsche, 'Homer's Contest'.)

123 the Islamic Renaissance radiating from Baghdad: Shenk, *The Immortal Game*, pp. 29–38.

123 In the twenty-first century, the United States is home to eleven of the fifteen top-rated universities in the world: *US News & World Report*, 'World's Best Colleges and Universities'.

124 New Haven pizza.

Principally Sally's and Pepe's – a great Wooster Street rivalry going back to 1938. Map at http://bit.ly/iepEc.

124 In his study of the ancient Greeks, Nietzsche imagined Plato declaring, 'Only the contest made me a poet, a sophist, an orator!': Nietzsche, 'Homer's Contest'.

124 'The ancient Greeks turned competition into an institution on which they based the education of their citizens,' explains Olympic official Cleanthis Palaeologos: Palaeologos, 'Sport and the Games in Ancient Greek Society'.

Also: 'The ancient Greeks in many respects stood as symbols of our commonly shared potential for overcoming artistic, intellectual, and athletic mediocrity,' writes Chicago State University's Alexander Makedon. (Makedon, 'In Search of Excellence'.)

Makedon's seventeen reasons for Greek success:

Some of the reasons mentioned by those who examined Greek culture, then, include, first, *democracy*, where free speech and public criticism were openly practised; and a corresponding hatred for tyrannies or one-man-rule of all kinds. Second, *striving for excellence* by the public at large. This happened through the internalisation over the centuries of the heroic or 'aristocratic' ideal by the masses, in the classical sense of 'aristocratic' as the rule of the excellent. Third, a corresponding effort at *moral excellence*, including not only constantly inquiring which life is worth living, but also people practising what they preached. Fourth, *fighting graft and corruption* at all levels, with a corresponding internalisation over the centuries of certain basic civic values. For example, even the slightest infraction by someone entrusted with a public office may lead not only to his dismissal, but also to his exile from the city state. Fifth, trying to *overcome personal weaknesses*, which may be seen as a corollary to their unusually intense attempts to excel. Sixth, behaving with the *highest integrity* even in the absence of immediate supervision. Seventh, subscribing to the 'agonistic' or *competitive spirit*, mostly through playful contests and competitions. Eighth, rewarding individuals on the *basis of merit*, as opposed to wealth, or family or political connections. This led to the birth of the Olympic Games in Greece, which in ancient Greece included not only physical, but also literary, dramatic, and musical contests. Ninth, instituting *education through play*. Tenth, designing a *whole city as the school*, by building it for personal effort and refinement, than mere protection from the elements. Eleventh, making public facilities *free to the poor*, so everyone could abundantly benefit from opportunities for self-development. Twelfth, *inviting young people to adult events*, where there were ample opportunities for learning through emulation by the young. In such situations, adults usually acted uprightly in their capacity as role models. Thirteenth, exercising *neighbourhood supervision* over the young, similar to the supervision exercised in Philippine barangays, except with many more opportunities for the worthy canalisation of youthful energy through sports, and artistic and educational contests. Fourteenth, the *institutionalisation through art of numerous role models*, including lining streets with statues of heroes. Fifteenth, involving numerous adults in a *city-wide network of mentors* who were not only unpaid, but considered it their honour to pay themselves for the pedagogical expenses of their protégés. Sixteenth, subscribing to an informal educational system of *expert itinerant teachers*, called 'sophists', who provided both an excellent education, and a model of excellence in learning, and were amply rewarded for their professional services. And seventeenth, placing a priority on *public service and philanthropy*, as contrasted to personal accumulation of wealth at the expense of the common good. For example, the wealthy were expected to pay a large part of the cost of large public projects. (Makedon, 'In Search of Excellence'.)

124 'Agonism implies a deep respect and concern for the other': Chambers, 'Language and Politics: Agonistic Discourse in the *West Wing*'.

124 Dutch historian Johan Huizinga suggests that without the agonistic spirit, human beings would simply be incapable of rising above mediocrity.

Alexander Makedon writes:

> Johan Huizinga formalised the cultural impact of play activities in his book *Homo Ludens: A Study of the Play Element in Culture.* The terms 'Homo Ludens' in Latin mean 'Man the Player'. His choice of words for a title contrasts with the traditional view of modern humans as 'homo sapiens', or man the thinker, perhaps to underline the priority that Huizinga assigned to the play element in the genesis of civilisation. According to Huizinga, great 'cultural' achievements are based on the agonistic or competitive spirit, without which humans would be at best 'mediocre'. As people compete for first place, they simultaneously force themselves to improve their skills, thus in the end reaching a higher plateau of educational achievement. Just as an impending athletic event forces athletes to prepare by intensifying their training, so are people striving to win [sic] finally achieve excellence. This is even more true when a whole culture adopts the agonistic or 'competitive' spirit, instead of merely a few institutions within that culture. (Makedon, 'In Search of Excellence'; Huizinga, *Homo Ludens*.)

125 Leonardo, Michelangelo, Raphael, Titian, and Correggio were all open-eyed adversaries: Goffen, *Renaissance Rivals*.

125 As soon as Florence began to build a new colossal *duomo*: Goffen, *Renaissance Rivals*, p. 7.

125 In fact, the Italian Renaissance actually began with a specific contest, according to Rutgers art historian Rona Goffen.

'The Renaissance was an inherently rivalrous age that began with a competition,' writes Rutgers historian Rona Goffen. 'Rivalry was institutionalised.' (Goffen, *Renaissance Rivals*, p. 4.)

125 The contest winner, Lorenzo Ghiberti: Goffen, *Renaissance Rivals*, pp. 4–7.

125 ***combattitori.***

My favourite word in this book – maybe my favourite word ever.

125 commissioned Leonardo and Michelangelo to work literally side by side: Anuar, 'Leonardo vs. Michelangelo'.

125 'Artists have always borrowed from each other,' writes Goffen. 'What is different about the sixteenth century is that the great masters . . . often knew each other's major patrons; and they knew each other, sometimes as friends and colleagues, sometimes as enemies – but always as rivals': Goffen, *Renaissance Rivals*, p. 26.

Also: 'The intentions to surpass one's rivals, past and present, distinguishes the Renaissance from earlier periods.' (Goffen, *Renaissance Rivals*, p. 3.)

126 At the time of its inception, though, Michelangelo was convinced that his commission from Pope Julius II: Goffen, *Renaissance Rivals*, pp. 215–16.

126 In 2006, economists Uri Gneezy, Kenneth L. Leonard, and John A. List compared competitive instincts in two very different societies: Maasai in Tanzania and Khasi in India. Among the patriarchal Maasai, men choose to compete at twice the rate of women. But among the Khasi, which is rooted in a matrilineal culture where women inherit property and children are named from the mother's side of the family, women choose to compete much more often than men.

Gneezy, Leonard, and List write:

> We observe some interesting data patterns. For example, Maasai men opt to compete at roughly twice the rate as Maasai women, evidence that is consistent with data from Western societies that use different tasks. Yet, this data pattern is reversed amongst the Khasi, where women choose the competitive environment considerably more often than Khasi men. We interpret these results as providing initial insights into the determinants of the observed gender differences. Viewed through the lens of our modelling framework, our results have import within the policy

community. For example, policymakers often are searching for efficient means to reduce the gender gap. If the difference in reaction to competition is based primarily on nature, then some might advocate, for example, reducing the competitiveness of the education system and labour markets in order to provide women with more chances to succeed. If the difference is based on nurture, or an interaction between nature and nurture, on the other hand, the public policy might be targeting the socialisation and education at early ages as well as later in life to eliminate this asymmetric treatment of men and women with respect to competitiveness. (Gneezy, Leonard, and List, 'Gender Differences in Competition: The Role of Socialization'.)

127 **If short-term tasks can be made relevant to long-term goals, researchers have found, then even LAMs will dive in and relish the challenge:** Tauer and Harackiewicz, 'Winning isn't everything', pp. 209–38; Durik and Harackiewicz, 'Achievement goals and intrinsic motivation', pp. 378–85.

128 **Mighton's response to this problem was to break down mathematical concepts into the most easily digestible form and help students build skills and confidence in tandem.**

An excerpt from *The Myth of Ability* demonstrates Mighton's approach:

F-1 Counting

First check whether your student can count on one hand by twos, threes, and fives. If they can't, you will have to teach them. I've found the best way to do this is to draw a hand like this:

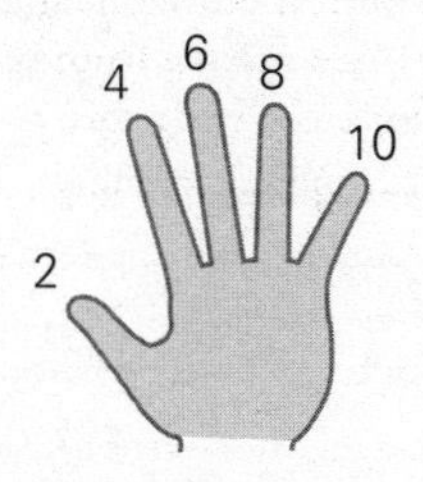

Have your student practise for a minute or two with the diagram, then without. When your student can count by twos, threes, and fives, teach them to multiply using their fingers, as follows:

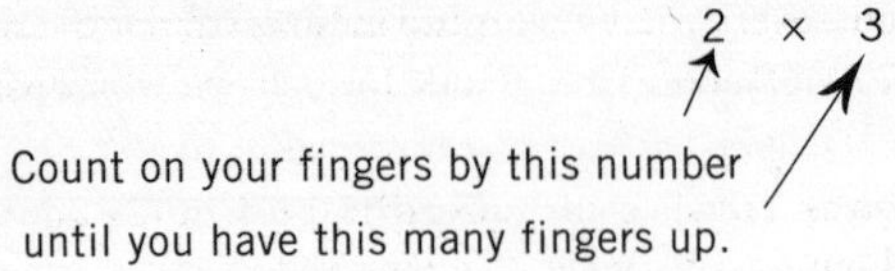

2 × 3

Count on your fingers by this number
until you have this many fingers up.

The number you reach is the answer.
Give your student practice with questions like:

4 x 5 = ____

2 x 3 = ____

3 x 3 = ____

3 x 5 = ____

5 x 2 = ____

Point out that 2 x 3 means: add three, two times (that's what you are doing as you count up on your fingers). Don't belabour this point, though – you can explain it in more depth when your student is further into the units. (Mighton, *The Myth of Ability*, pp. 64–65.)

128 **'With proper teaching and minimal tutorial support':** Mighton, *The Myth of Ability*, p. 21.

128 **Mighton does not claim his particular teaching method as the only approach, or even the best:** Mighton, *The Myth of Ability*, p. 27.
—⟫ John Mighton is also an actor who played a prominent role in the movie *Good Will Hunting*. The irony is that the message of the movie – brilliance is innate – runs counter to his marvellous work at JUMP.

128 **In fact, countless students fall behind in maths and other subjects for exactly the same reason others generally hate to compete directly in any field:** Tauer and Harackiewicz, 'Winning isn't everything', pp. 209–38; Durik and Harackiewicz, 'Achievement goals and intrinsic motivation', pp. 378–85.

128 **'I wasn't quite suited for the educational system,' Bruce Springsteen has said:** Interview conducted by Ted Koppel, ABC's *Nightline Up Close*.

128 **'If non-linear leaps in intelligence and ability are possible':** Mighton, *The Myth of Ability*, p. 19.

129 'Man – every man – is an end in himself, not the means to the ends of others,' Ayn Rand wrote: Rand, 'Introducing Objectivism'.

129 'Kenyan coaches can afford to push their athletes to the most extreme boundaries': Wolff, 'No Finish Line'.

CHAPTER 10:
GENES 2.1 – HOW TO IMPROVE YOUR GENES

PRIMARY SOURCES

Harper, Lawrence V. 'Epigenetic inheritance and the intergenerational transfer of experience'. *Psychological Bulletin* 131, no. 3 (2005): 340–60.

Jablonka, Eva, and Marion J. Lamb. *Evolution in Four Dimensions*. MIT Press, 2005.

Morgan, Hugh D., Heidi G. E. Sutherland, David I. K. Martin, and Emma Whitelaw. 'Epigenetic inheritance at the agouti locus in the mouse'. *Nature Genetics* 23 (1999): 314–18.

Watters, Ethan. 'DNA Is Not Destiny'. Published on the *Discover* Web site, November 22, 2006. (A superb piece, without which I would have been unable to write this chapter.)

CHAPTER NOTES

130 In textbooks and elsewhere, Lamarckism has been defined (and mocked) as a crude, pre-Darwinian conception of evolution, tainted by the flimsy idea that biological heredity can somehow be altered through personal experience.

An important corrective of the Lamarck legacy, from Eva Jablonka and Marion Lamb:

> This often repeated version of the history of evolutionary ideas is wrong in many respects: it is wrong in making Lamarck's ideas seem so simplistic, wrong in implying that Lamarck invented the idea that acquired characteristics are inherited, wrong in not recognising that use and disuse had a place in Darwin's thinking too, and wrong to suggest that the theory of natural selection displaced the inheritance of acquired characters from the mainstream of evolutionary thought. The truth is that Lamarck's theory was quite sophisticated, encompassing much more than the inheritance of acquired characters. Moreover, Lamarck did not invent the idea that acquired characters can be inherited – almost all biologists believed this at the beginning of the nineteenth century,

and many still believed it at its end. (Jablonka and Lamb, *Evolution in Four Dimensions*, p. 13; see also Ghiselin, 'The Imaginary Lamarck: A Look at Bogus "History" in Schoolbooks'.)

130 Lamarck called it 'the inheritance of acquired characteristics' – the notion that an individual's actions can alter the biological inheritance passed on to his or her children.

Lamarck wrote:

> All the acquisitions or losses wrought by nature on individuals, through the influence of the environment in which their race has long been placed, and hence through the influence of the predominant use or permanent disuse of any organ; all these are preserved by reproduction to the new individuals which arise, provided that the acquired modifications are common to both sexes, or at least to the individuals which produce the young. (Lamarck, *Zoological Philosophy*, p. 113.)

130 For example, giraffes, according to Lamarck's theory, had developed longer and longer necks over the generations because of the giraffe's practice of reaching higher and higher for food.

Lamarck wrote:

> It is interesting to observe the result of habit in the peculiar shape and size of the giraffe: this animal, the tallest of the mammals, is known to live in the interior of Africa in places where the soil is nearly always arid and barren, so that it is obliged to browse on the leaves of trees and to make constant efforts to reach them. From this habit long maintained in all its race, it has resulted that the animal's forelegs have become longer than its hind-legs, and that its neck is lengthened to such a degree that the giraffe, without standing up on its hind-legs, attains a height of six metres. (Lamarck, *Philosophie Zoologique*, as quoted in Gould, *The Structure of Evolutionary Theory*, p. 188.)

131 ***Drawing of Giraffe in a 'classic' feeding position, extending its neck, head, and tongue to reach the leaves of an Acacia tree. Tsavo National Park, Kenya*:** Drawing by C. Holdrege. (Holdrege, *In Context* #10, pp. 14–19.)

131 After Darwin's *Origin of Species* and the subsequent discovery of genes, a very different notion – the theory of natural selection – became scientific and popular consensus.

—◦⪢ Actually, what the general public still refers to as our 'Darwinian' understanding of evolution is more properly called the 'modern evolutionary synthesis', a melding of Darwin's ideas with later genetics discoveries.

Here is a nice synopsis of the modern evolutionary synthesis, from Douglas J. Futuyma:

> The major tenets of the evolutionary synthesis, then, were that populations contain genetic variation that arises by random (i.e. not adaptively directed) mutation and recombination; that populations evolve by changes in gene frequency brought about by random genetic drift, gene flow, and especially natural selection; that most adaptive genetic variants have individually slight phenotypic effects so that phenotypic changes are gradual (although some alleles with discrete effects may be advantageous, as in certain colour polymorphisms); that diversification comes about by speciation, which normally entails the gradual evolution of reproductive isolation among populations; and that these processes, continued for sufficiently long, give rise to changes of such great magnitude as to warrant the designation of higher taxonomic levels (genera, families, and so forth). (Futuyma, *Evolutionary Biology*, p. 12.)

132 **Pictures of Toadflax flowers:** Emil Nilsson. Used by permission.

132 **There *was* a difference between the two flowers on their respective epigenomes:** Jablonka and Lamb, *Evolution in Four Dimensions*, p. 142.

132 **DNA is famously wound together in a double-helix strand.**

Diameter of DNA is about 20 angstroms (1 angstrom = 1×10^{-10} metres).

133 **These histones protect the DNA and keep it compact. They also serve as a mediator for gene expression, telling genes when to turn on and off. It's been known for many years that this epigenome ('epi-' is a Latin prefix for 'above' or 'outside') can be altered by the environment and is therefore an important mechanism for gene-environment interaction.**

'In 2005, Madrid biologist Manel Esteller and colleagues reported finding significant epigenetic differences in a whopping thirty-five per cent of identical twin sets. "These findings help show how environmental factors can change one's gene expression and susceptibility to disease," said Esteller.' (Choi, 'How Epigenetics Affects Twins'; see also Pray, 'Epigenetics', pp. 1, 4.)

133 **They observed that their batch of genetically identical mice were turning up with a range of different fur colours:** Morgan, Sutherland, Martin, and Whitelaw, 'Epigenetic inheritance at the agouti locus in the mouse', pp. 314–18.

133 **A pregnant yellow mouse eating a diet rich in folic acid or soy milk would be prone to experience an epigenetic mutation producing brown-fur offspring, and even with the pups returning to a normal diet, that brown fur would be passed to future generations.**

Morgan and Whitelaw write:

> Changes to the dam's diet during pregnancy can alter the proportion of yellow mice within a litter. For example, when the dam's diet is supplemented with methyl donors, including betaine, methionine, and folic acid, there is a shift in the colour of their offspring away from yellow and towards agouti. Similar effects have been observed following the feeding of the dams with genistein, which is found in soy milk. (Morgan and Whitelaw, 'The case for transgenerational epigenetic inheritance in humans', pp. 394–95.)

134 **exposure to a pesticide in one generation of rats spurred an epigenetic change:** Watters, 'DNA Is Not Destiny'.

134 **age-related epigenetic changes in human males:** Malaspina et al., 'Paternal age and intelligence', pp. 117–25.

134 **nutritional deficiencies and cigarette smoking in one generation of humans had effects across several generations:** Watters, 'DNA Is Not Destiny'.

134 **link between inherited epigenetic changes and human colon cancer:** Hitchins et al., 'Inheritance of a cancer-associated MLH1 germ-line epimutation', pp. 697–705.

134 **'Epigenetics is proving we have some responsibility for the integrity of our genome,' says the Director of Epigenetics and Imprinting at Duke University, Randy Jirtle:** Watters, 'DNA Is Not Destiny'.

135 **'Information is transferred from one generation to the next by many interacting inheritance systems':** Jablonka and Lamb, *Evolution in Four Dimensions*, p. 319.

135 New animal research in the February 4 [2009] issue of *The Journal of Neuroscience* shows that a stimulating environment improved the memory of young mice with a memory-impairing genetic defect and also improved the memory of their eventual offspring: Society for Neuroscience, 'Mother's Experience Impacts Offspring's Memory'; the original article cited is Arai, Li, Hartley, and Feig, 'Transgenerational rescue of a genetic defect in long-term potentiation and memory formation by juvenile enrichment', pp. 1496–1502.

136 'People used to think that once your epigenetic code was laid down in early development, that was it for life,' says McGill University epigenetics pioneer Moshe Szyf: Watters, 'DNA Is Not Destiny'.

EPILOGUE: TED WILLIAMS FIELD

137 His tiny boyhood home at 4121 Utah Street still stands.
http://bit.ly/9Bmml.

137 Two short blocks away, his old practice baseball field is still there too.

http://bit.ly/yUGZs.

Bibliography

Note: For a digital version of this bibliography, complete with source links, visit geniusbibliography.davidshenk.com.

Abrams, Michael. 'The Biology of . . . Perfect Pitch: Can Your Child Learn Some of Mozart's Magic?' *Discover*, December 1, 2001.

American Institute of Physics. 'Slam Dunk Science: Physicist Explains Basic Principles Governing Basketball'. November 1, 2007.

American Psychological Association. 'Intelligence: Knowns and Unknowns. Report of a Task Force Established by the Board of Scientific Affairs of the American Psychological Association'. Released August 7, 1995.

———. 'Intelligence: knowns and unknowns'. *American Psychologist* 51, no. 2 (February 1996): 77–101.

Andersen, J. L., H. Klitgaard, and B. Saltin. 'Myosin heavy chain isoforms in single fibres from m. vastus lateralis of sprinters: influence of training'. *Acta Physiologica Scandinavica* 151 (1994): 135–42.

Anderson, Jesper L., Peter Schjerling, and Bengt Saltin. 'Muscle, Genes and Athletic Performance'. *Scientific American*, September 2000.

Anderson, John R. *Cognitive Skills and Their Acquisition.* Lawrence Erlbaum, 1981.

Angier, Natalie. 'Separated by Birth?' *New York Times*, February 8, 1998.

Anuar, A. H. 'Leonardo vs. Michelangelo: The Battle Between the Masters'. Published on the Holiday City Web site, November 30, 2004.

Arai, J., S. Li, D. M. Hartley, and L. A. Feig. 'Transgenerational rescue of a genetic defect in long-term potentiation and memory formation by juvenile enrichment'. *The Journal of Neuroscience* 29, no. 5 (February 4, 2009): 1496–1502.

Baker, Catherine. Report on Eric Turkheimer's presentation 'Three Laws of Behavior Genetics and What They Mean'. *Program of Dialogue on Science, Ethics, & Religion*, April 10, 2003; published on the American Association for the Advancement of Science Web site.

Bale, John. Comment on *The Sports Factor* radio show, February 28, 1997.

———. *Sports Geography*. Routledge, 2003.

Baltes, Paul B. 'Testing the limits of the ontogenetic sources of talent and excellence'. *Behavioral and Brain Sciences* 21, no. 3 (June 1998): 407–8.

Bamberger, J. 'Growing Up Prodigies: The Mid-life Crisis'. In *Developmental Approaches to Giftedness and Creativity*, edited by D. H. Feldman. Jossey-Bass, 1982, pp. 61–67.

Bannister, R. G. 'Muscular effort'. *British Medical Bulletin* 12 (1956): 222–25.

Barlow, F. *Mental Prodigies*. Greenwood Press, 1952.

Bate, Karen. '"Dora the Explorer" Shows Pupils the Way'. *Salisbury Journal*, September 30, 2006.

Bateson, Patrick. 'Behavioral Development and Darwinian Evolution'. In *Cycles of Contingency: Developmental Systems and Evolution*, edited by Susan Oyama et al. MIT Press, 2003, pp. 149–66.

Bateson, Patrick, and Matteo Mameli. 'The innate and the acquired: useful clusters or a residual distinction from folk biology?' *Developmental Psychobiology* 49 (2007): 818–31.

Bateson, Patrick, and Paul Martin. *Design for a Life: How Biology and Psychology Shape Human Behavior*. Simon & Schuster, 2001.

Baumrind, D. 'Child care practices anteceding three patterns of preschool behavior'. *Genetic Psychology Monographs* 75 (1967): 43–88.

Benard, Bonnie. *Resiliency: What We Have Learned*. WestEd, 2004.

Benn-Torres, J., et al. 'Admixture and population stratification in African Caribbean populations'. *Annals of Human Genetics* 72, no. 1 (2008): 90–98.

Berg, Kate, et al. (The Race, Ethnicity, and Genetics Working Group of the National Human Genome Research Institute). 'The use of racial, ethnic, and ancestral categories in human genetics research'. *American Journal of Human Genetics* 77, no. 4 (October 2005): 519–32.

Bilger, Burkhard. 'The Height Gap: Why Europeans Are Getting Taller and Taller – and Americans Aren't'. *New Yorker*, April 5, 2004.

Binet, Alfred. *Mnemonic Virtuosity: A Study of Chess Players*. 1893. Translated by Marianne L. Simmel and Susan B. Barron. Journal Press, 1966.

———. *Les idées modernes sur les enfants* (Modern Ideas on Children). Flammarion, 1909. Reprinted in 1973.

Birbaumer, N. 'Rain Man's revelations'. *Nature* 399 (1999): 211–12.

Blackwell, Lisa S., Kali H. Trzesniewski, and Carol Sorich Dweck. 'Implicit theories of intelligence predict achievement across an adolescent transition: a longitudinal study and an intervention'. *Child Development* 78, no. 1 (January/February 2007): 246–63.

Bloom, B. *Developing Talent in Young People*. Ballantine, 1985.

Bloom, Marc. 'Kenyan Runners in the U.S. Find Bitter Taste of Success'. *New York Times*, April 16, 1998.

Bottinelli, Roberto, and Carlo Reggiani, eds. *Skeletal Muscle Plasticity in Health and Disease*. Springer, 2006.

Bouchard, T. J., and M. McGue. 'Familial studies of intelligence: a review'. *Science* 212, no. 4498 (1981): 1055–59.

———. 'Genetic and environmental influences on human psychological differences'. *Journal of Neurobiology* 54 (2003): 4–45.

Bradshaw, E., and M. A. McHenry. 'Pitch discrimination and pitch matching abilities of adults who sing inaccurately'. *Journal of Voice* 19, no. 3 (September 2005): 431–39.

Brazelton, T. Berry. *Touchpoints: Your Child's Emotional and Behavioral Development, Birth to 3*. Capo Lifelong Books, 1992.

Brockman, John. 'Design for a Life: A Talk with Patrick Bateson'. EDGE 67, April 23, 2000.

Bronson, Po. 'How Not to Talk to Your Kids: The Inverse Power of Praise'. *New York*, February 12, 2007.

Brown, Kathryn. 'Striking the Right Note'. *New Scientist*, December 4, 1999.

Bruer, J. *The Myth of the First Three Years*. Free Press, 1999.

Brutsaert, Tom D., and Esteban J. Parra. 'What makes a champion? Explaining variation in human athletic performance'. *Respiratory Physiology and Neurobiology* 151 (2006): 109–23.

Budgett, R. 'ABC of sports medicine: the overtraining syndrome'. *British Medical Journal* 309 (1994): 465–68.

Burke, Ed. *High-Tech Cycling*. Human Kinetics, 2003.

Campitelli, G., and F. Gobet. 'The role of practice in chess: a longitudinal study'. *Learning and Individual Differences* 18, no. 4 (2008): 446–58.

Ceci, S. J. *On Intelligence: A Bio-ecological Treatise on Intellectual Development*. Harvard University Press, 1996.

Ceci, S. J., T. Rosenblum, E. de Bruyn, and D. Lee. 'A Bio-Ecological Model of Intellectual Development: Moving Beyond h2'. In *Intelligence, Heredity, and Environment*, edited by R. J. Sternberg and E. Grigorenko. Cambridge University Press, 1997, pp. 303–22.

Chambers, Samuel. 'Language and Politics: Agonistic Discourse in *The West Wing*'. Published on Ctheory.net, an online journal edited by Arthur Kroker and Marilouise Kroker, November 12, 2001.

Charness, Neil, R. Th. Krampe, and U. Mayr. 'The Role of Practice and Coaching in Entrepreneurial Skill Domains: An International Comparison of Life-Span Chess Skill Acquisition'. In *The Road to Excellence: The Acquisition of Expert Performance in the Arts and Sciences, Sports, and Games*, edited by K. A. Ericsson. Lawrence Erlbaum, 1996, pp. 51–80.

Charness, Neil, M. Tuffiash, R. Krampe, E. Reingold, and E. Vasyukova. 'The role of deliberate practice in chess expertise'. *Applied Cognitive Psychology* 19 (2005): 151–65.

Chase, David, and Terence Winter. 'The Sopranos: Walk Like a Man'. Season 6, episode 17. Original air date May 6, 2007.

Chase, W. G., and H. A. Simon. 'The Mind's Eye in Chess'. *Visual Information Processing: Proceedings of the 8th Annual Carnegie Psychology Symposium.* Academic Press, 1972.

Chen, Edwin. 'Twins Reared Apart: A Living Lab'. *New York Times Magazine*, December 9, 1979.

Choi, Charles Q. 'How Epigenetics Affects Twins'. *News from The Scientist*, July 7, 2005.

Clark, Matthew. 'How Tiny Jamaica Develops So Many Champion Sprinters'. *Christian Science Monitor*, June 27, 2008.

Clarke, Ann M., and Alan D. Clarke. *Early Experience and the Life Path.* Somerset, 1976.

Coetzer, P., T. D. Noakes, B. Sanders, M. I. Lambert, A. N. Bosch, T. Wiggins, and S. C. Dennis. 'Superior fatigue resistance of elite black South African distance runners'. *Journal of Applied Physiology* 75 (1993): 1822–27.

Colangelo, N., S. Assouline, B. Kerr, R. Huesman, and D. Johnson. 'Mechanical Inventiveness: A Three-Phase Study'. In *The Origins and Development of High Ability*, edited by G. R. Bock and K. Ackrill. Wiley, 1993, pp. 160–74.

Crabbe, John C., Douglas Wahlsten, and Bruce C. Dudek. 'Genetics of mouse behavior: interactions with laboratory environment'. *Science* 284, no. 5420 (June 4, 1999): 1670–72.

Cravens, H. 'A scientific project locked in time: the Terman Genetic Studies of Genius'. *American Psychologist* 47, no. 2 (February 1992): 183–89.

Csikszentmihályi, M., and I. S. Csikszentmihályi. 'Family influences on the development of giftedness'. *Ciba Foundation Symposium* 178 (1993): 187–200.

Csikszentmihályi, M., Kevin Rathunde, and Samuel Whalen. *Talented Teenagers: The Roots of Success and Failure.* Cambridge University Press, 1993.

Dalla Bella, Simone, Jean-François Giguère, and Isabelle Peretz. 'Singing proficiency in the general population'. *Journal of the Acoustical Society of America* 1212 (February 2007): 1182–89.

Deary, Ian J., Martin Lawn, and David J. Bartholomew. 'A conversation between Charles Spearman, Godfrey Thomson, and Edward L. Thorndike: The International Examinations Inquiry Meetings, 1931–1938'. *History of Psychology* 11, no. 2 (May 2008): 122–42.

de Groot, Adrianus Dingeman. *Thought and Choice in Chess*. Walter de Gruyter, 1978.

Denison, Niki. 'The Rain Man in All of Us'. *On Wisconsin*, Summer 2007.

Deutsch, Diana. 'Tone Language Speakers Possess Absolute Pitch'. Presentation at the 138th meeting of the Acoustical Society of America, November 4, 1999.

De Vany, Art. 'Twins'. Published on his blog, September 9, 2005.

Diamond, M., and J. L. Hopson. *Magic Trees of the Mind: How to Nurture Your Child's Intelligence, Creativity, and Healthy Emotions from Birth Through Adolescence*. Penguin, 1999.

Dickens, William T., and James R. Flynn. 'Heritability estimates versus large environmental effects: the IQ paradox resolved'. *Psychological Review* 108, no. 2 (2001): 346–69.

Dickinson, Amy. 'Little Musicians'. *Time*, December 13, 1999.

Dingfelder, S. 'Most people show elements of absolute pitch'. *Monitor on Psychology* 36, no. 2 (February 2005): 33.

Dinwiddy, John Rowland. *Bentham*. Oxford University Press, 1989.

Dodge, Kenneth A. 'The nature-nurture debate and public policy'. *Merrill-Palmer Quarterly* 50, no. 4 (2004): 418–27.

Dornbusch, Sanford M., Philip L. Ritter, P. Herbert Leiderman, Donald F. Roberts, and Michael J. Fraleigh. 'The relation of parenting style to adolescent school performance'. *Child Development* 58, no. 5 (October 1987): 1244–57.

Downes, Stephen M. 'Heredity and Heritability'. Published online on the Stanford Encyclopedia of Philosophy Web site, first posted July 15, 2004; revised May 28, 2009.

Duffy, D. L., et al. 'A three-single-nucleotide polymorphism haplotype in intron 1 of OCA2 explains most human eye-color variation'. *American Journal of Human Genetics* 80, no. 2 (February 2007): 241–52.

Duffy, L. J., B. Baluch, and K. A. Ericsson. 'Dart performance as a function of facets of practice amongst professional and amateur men and women players'. *International Journal of Sport Psychology* 35 (2004): 232–45.

Durik, Amanda M., and Judith M. Harackiewicz. 'Achievement goals and intrinsic motivation: coherence, concordance, and achievement orientation'. *Journal of Experimental Social Psychology* 39, no. 4 (2003): 378–85.

Dweck, Carol. *Mindset: The New Psychology of Success*. Random House, 2006.

Edes, Gordon. 'Gone: In Baseball and Beyond, Williams Was a True American Hero'. *Boston Globe*, July 6, 2002.

Edmonds, R. 'Characteristics of Effective Schools'. In *The School Achievement of Minority Children: New Perspectives*, edited by U. Neisser. Lawrence Erlbaum, 1986, pp. 93–104.

Einstein, Alfred. Preface to Mozart, Leopold. *A Treatise on the Fundamental Principles of Violin Playing*. Oxford University Press, 1985.

Eisenberg, Leon. 'Nature, niche, and nurture: the role of social experience in transforming genotype into phenotype'. *Academic Psychiatry* 22 (December 1998): 213–22.

Elbert, Thomas, Christo Pantev, Christian Wienbruch, Brigitte Rockstroh, and Edward Taub. 'Increased cortical representation of the fingers of the left hand in string players'. *Science* 270 (1995): 305–7.

Elliot, Andrew J., and Carol S. Dweck, eds. *Handbook of Competence and Motivation*. Guilford Publications, 2005.

Entine, Jon. *Taboo: Why Black Athletes Dominate Sports and Why We Are Afraid to Talk About It*. Public Affairs, 2000.

———. 'Jewish Hoop Dreams: 1920s and '30s Ghetto Jews Transformed the Game'. *Jewish News of Greater Phoenix*, June 22, 2001.

Ericsson, K. Anders. 'Superior memory of experts and long-term working memory'. Updated and extracted version. Published on the Florida State University Department of Psychology Web site, http://www.psy.fsu.edu// faculty/ericsson/ ericsson.mem.exp.html.

———. 'Deliberate practice and the modifiability of body and mind: toward a science of the structure and acquisition of expert and elite performance'. *International Journal of Sport Psychology* 38 (2007): 4–34.

Ericsson, K. Anders, and Neil Charness. 'Expert performance – its structure and acquisition'. *American Psychologist*, August 1994.

Ericsson, K. Anders, Neil Charness, Paul J. Feltovich, and Robert R. Hoffman, eds. *The Cambridge Handbook of Expertise and Expert Performance*. Cambridge University Press, 2006.

Ericsson, K. A., W. G. Chase, and S. Faloon. 'Acquisition of a memory skill'. *Science* 208 (1980): 1181–82.

Ericsson, K. A., and W. Kintsch. 'Long-term working memory'. *Psychological Review* 102, no. 2 (1995): 211–45.

Ericsson, K. Anders, Roy W. Roring, and Kiruthiga Nandagopal. 'Giftedness and evidence for reproducibly superior performance: an account based on the expert performance framework'. *High Ability Studies* 18, no. 1 (June 2007): 3–56.

Farber, S. L. *Identical Twins Reared Apart: A Reanalysis*. New York: Basic Books, 1981.

Farrey, Tom. 'Awaiting Another Chip off Ted Williams' Old DNA?' Published on the ESPN.com Web site, http://espn.go.com/gen/s/2002/0709/1403734.html, July 9, 2002.

Feist, Gregory J. 'The Evolved Fluid Specificity of Human Creative Talent'. In *Creativity: From Potential to Realization*, edited by R. J. Sternberg, E. L. Grigorenko, and J. L. Singer. American Psychological Association, 2004, pp. 57–82.

Feldman, David Henry. 'A follow-up of subjects scoring above 180 IQ in Terman's genetic studies of genius'. *Council for Exceptional Children* 50, no. 6 (1984): 518–23.

Fest, Sebastian. '"Actinen A": Jamaica's Secret Weapon'. Deutsche Presse-Agentur, August 14, 2008.

Field Museum. Gregor Mendel: Planting the Seeds of Genetics. Exhibition, September 15, 2006–April 1, 2007. http://www.fieldmuseum.org/mendel/story_pea.asp.

Flynn, J. R. 'Massive IQ gains in 14 nations: what IQ tests really measure'. *Psychological Bulletin* 101 (1987): 171–91.

———. 'Beyond the Flynn Effect: Solution to All Outstanding Problems Except Enhancing Wisdom'. Lecture at the Psychometrics Centre, Cambridge Assessment Group. University of Cambridge, December 16, 2006.

———. *What Is Intelligence? Beyond the Flynn Effect*. Cambridge University Press, 2007.

Fox, Paul W., Scott L. Hershberger, and Thomas J. Bouchard Jr. 'Genetic and environmental contributions to the acquisition of a motor skill'. *Nature* 384 (1996): 356.

Freeman, J. 'Families, the Essential Context for Gifts and Talents'. In *International Handbook of Research and Development of Giftedness and Talent*, edited by K. A. Heller, F. J. Monks, R. Sternberg, and R. Subotnik. Pergamon Press, 2000, pp. 573–85.

———. 'Teaching for Talent: Lessons from the Research'. In *Developing Talent Across the Lifespan*, edited by C. F. M. Lieshout and P. G. Heymans. Psychology Press, 2000, pp. 231–48.

———. 'Giftedness in the long term'. *Journal for the Education of the Gifted* 29 (2006): 384–403.

Friend, Tim. 'Blueprint for Life'. *USA Today*, January 26, 2003.

Frudakis, T., T. Terravainen, and M. Thomas. 'Multilocus OCA2 genotypes specify human iris colors'. *Human Genetics* 122, no. 3/4 (November 2007): 311–26.

Futuyma, D. J. *Evolutionary Biology*. Sinauer Associates, 1986.

Galton, Francis. *Hereditary Genius*. MacMillan, 1869.

———. *English Men of Science: Their Nature and Nurture*. D. Appleton, 1874.

Galton, Francis, and Charles Darwin. Correspondence published on the Galton.org Web site.

Gardner, H. *Creating Minds: An Anatomy of Creativity Seen Through the Lives of Freud, Einstein, Picasso, Stravinsky, Eliot, Graham and Gandhi*. Basic Books, 1993.

Gardner, Howard. 'Do Parents Count?' *New York Review of Books*, November 5, 1998.

Gardner, Howard. *Intelligence Reframed: Multiple Intelligences for the 21st Century*. Basic Books, 1999.

Garfield, E. 'High Impact Science and the Case of Arthur Jensen'. In *Essays of an Information Scientist*, vol. 3, 1977–78, pp. 652–62. Current Contents 41, pp. 652–62, October 9, 1978.

Gazzaniga, Michael S. 'Smarter on Drugs'. *Scientific American Mind*, October 2005.

Geiringer, Karl. 'Leopold Mozart'. *The Musical Times* 78, no. 1131 (May 1937): 401–4.

Ghiselin, Michael T. 'The Imaginary Lamarck: A Look at Bogus "History" in Schoolbooks'. *The Textbook Letter*, September/October 1994.

Gladwell, Malcolm. 'Kenyan Runners'. Published on the Gladwell.com Web site, November 16, 2007.

Gneezy, Uri, Kenneth L. Leonard, and John A. List. 'Gender Differences in Competition: The Role of Socialization'. UCSB Seminar Paper. Published on the University of California, Santa Barbara, Department of Economics Web site, June 19, 2006.

Gobet, F., and G. Campitelli. 'The role of domain-specific practice, handedness and starting age in chess'. *Developmental Psychology* 43 (2007): 159–72.

Godfrey-Smith, Peter. 'Genes and Codes: Lessons from the Philosophy of Mind?' In *Biology Meets Psychology: Constraints, Conjectures, Connections*, edited by V. Q. Hardcastle. MIT Press, 1999, 305–31.

Goffen, Rona. *Renaissance Rivals: Michelangelo, Leonardo, Raphael, Titian*. Yale University Press, 2004.

Goleman, Daniel. *Destructive Emotions: A Scientific Dialogue with the Dalai Lama*. Bantam, 2003.

Gordon, H. W. 'Hemisphere asymmetry in the perception of musical chords'. *Cortex* 6 (1970): 387–98.

———. 'Left-hemisphere dominance of rhythmic elements in dichotically presented melodies'. *Cortex* 14 (1978): 58–70.

———. 'Degree of ear asymmetry for perception of dichotic chords and for illusory chord localization in musicians of different levels of competence'. *Journal of Experimental Psychology: Perception and Performance* 6 (1980): 516–27.

Gottlieb, Gilbert. 'On making behavioral genetics truly developmental'. *Human Development* 46 (2003): 337–55.

Gould, Stephen Jay. *The Mismeasure of Man*. Norton, 1996.

Grathoff, Pete. 'Science of Hang Time'. *The Kansas City Star*, November 29, 2008.

Green, Christopher D. Classics in the History of Psychology Web site.

Greulich, William Walter. 'A comparison of the physical growth and development of American-born and native Japanese children'. *American Journal of Physical Anthropology* 15 (1997): 489–515.

Griffiths, Paul. 'The Fearless Vampire Conservator: Phillip Kitcher and Genetic Determinism'. In *Genes in Development: Rereading the Molecular Paradigm*, edited by E. M. Neumann-Held and C. Rehmann-Sutter. Duke University Press, 2006.

Grigorenko, Elena. 'The relationship between academic and practical intelligence: a case study of the tacit knowledge of native American Yup'ik people in Alaska'. Office of Educational Research and Improvement, December 2001.

Gusnard, Debra A., et al. 'Persistence and brain circuitry'. *Proceedings of the National Academy of Sciences* 100, no. 6 (March 18, 2003): 3479–84.

Halberstam, David. *Playing for Keeps*. Broadway Books, 2000.

Hall, Wayne D., Katherine I. Morley, and Jayne C. Lucke. 'The prediction of disease risk in genomic medicine'. *EMBO Reports* 5, S1 (2004): S22–S26.

Hamilton, Bruce. 'East African running dominance: what is behind it?' *British Journal of Sports Medicine* 34 (2000): 391–94.

Harper, Lawrence V. 'Epigenetic inheritance and the intergenerational transfer of experience'. *Psychological Bulletin* 131, no. 3 (2005): 340–60.

Harris, Judith Rich. *The Nurture Assumption: Why Children Turn Out the Way They Do*. Simon & Schuster, 1999.

Hart, Betty, and Todd R. Risley. 'The early catastrophe: the 30 million word gap by age 3'. *American Educator* 27, no. 1 (2003).

Hassler, M. 'Functional cerebral asymmetric and cognitive abilities in musicians, painters, and controls'. *Brain and Cognition* 13 (1990): 1–17.

Hassler, M., and N. Birbaumer. 'Handedness, musical attributes, and dichaptic and dichotic performance in adolescents: a longitudinal study'. *Developmental Neuropsychology* 4, no. 2 (1988): 129–45.

Hattiangadi, Nina, Victoria Husted Medvec, and Thomas Gilovich. 'Failing to act: regrets of Terman's geniuses'. *International Journal of Aging and Human Development* 40, no. 3 (1995): 175–85.

Hays, Kristen. 'A Year Later, Cloned Cat Is No Copycat: Cc Illustrates the Complexities of Pet Cloning'. Associated Press, November 4, 2003.

Hermann, Evelyn. *Shinichi Suzuki: The Man and His Philosophy*. Senzay, 1981.

Herrnstein, Richard J., and Charles Murray. *The Bell Curve*. Free Press, 1994.

Hertzig, Margaret E., and Ellen A. Farber. *Annual Progress in Child Psychiatry and Child Development 1997*. Routledge, 2003.

Highfield, Roger. 'Unfaithful? I'm Sorry Darling but It's All in My Genes'. *Daily Telegraph*, November 25, 2004.

Hitchins, M. P., et al. 'Inheritance of a cancer-associated MLH1 germ-line epimutation'. *New England Journal of Medicine* 356 (2007): 697–705.

Holdrege, Craig. 'The giraffe's short neck'. *In Context* 10 (Fall 2003): 14–19.

Holt, Jim. 'Measure for Measure: The Strange Science of Francis Galton'. *New Yorker*, January 24–31, 2005.

Howe, Michael J. A. 'Can IQ Change?' *The Psychologist*, February 1998.

———. *Genius Explained*. Cambridge University Press, 1999.

Howe, Michael J. A., J. W. Davidson, and J. A. Sloboda. 'Innate talents: reality or myth'. *Behavioural and Brain Sciences* 21 (1998): 399–442.

Huizinga, Johan. *Homo Ludens: A Study of the Play Element in Culture*. Roy Publishers, 1950.

Hulbert, Ann. 'The Prodigy Puzzle'. *New York Times*, November 20, 2005.

Human Genome Project. 'How Many Genes Are in the Human Genome?' Published on the Oak Ridge National Laboratory Web site.

Humphrey, N. 'Comments on shamanism and cognitive evolution'. *Cambridge Archaeological Journal* 12, no. 1 (2002): 91–94.

Jablonka, Eva, and Marion J. Lamb. *Evolution in Four Dimensions*. MIT Press, 2005.

Johnson, Mark H., and Annette Karmiloff-Smith. 'Neuroscience Perspectives on Infant Development'. In *Theories of Infant Development*, edited by J. Gavin Bremmer and Alan Slater. Blackwell Publishing, 2003.

Johnston, Timothy D., and Laura Edwards. 'Genes, interactions, and the development of behavior'. *Psychological Review* 109, no. 1 (2002): 26–34.

Jones, H. E., and N. Bayley. 'The Berkeley Growth Study'. *Child Development* 12 (1941): 167–73.

Joseph, Jay. *The Gene Illusion: Genetic Research in Psychiatry and Psychology Under the Microscope*. Algora Publishing, 2004.

Kagan, J. *Three Seductive Ideas*. Harvard University Press, 1998.

Kalmus, H., and D. B. Fry. 'On tune deafness (dysmelodia): frequency, development, genetics and musical background'. *Annals of Human Genetics* 43, no. 4 (May 1980): 369–82.

Keirn, Jennifer. 'Who's in Charge? Teach Kids Self-Control'. *Family Magazine*, July 2007.

Keller, Evelyn Fox. *The Century of the Gene*. Harvard University Press, 2002.

Khoury, M. J., Q. Yang, M. Gwinn, J. Little, and W. D. Flanders. 'An epidemiological assessment of genomic profiling for measuring susceptibility to common diseases and targeting interventions'. *Genetics in Medicine* 6 (2004): 38–47.

Kliegl, Smith, and P. B. Baltes. 'On the locus and process of magnification of age differences during mnemonic training'. *Developmental Psychology* 26 (1990): 894–904.

Kohn, Tertius A., Birgitta Essén-Gustavsson, and Kathryn H. Myburgh. 'Do skeletal muscle phenotypic characteristics of Xhosa and Caucasian endurance runners differ when matched for training and racing distances?' *Journal of Applied Physiology* 103 (2007): 932–40.

Kolata, Gina. 'Identity: Just What Are Your Odds in Genetic Roulette? Go Figure'. *New York Times*, March 8, 2000.

Koppel, Ted. *Nightline Up Close*. ABC, aired July 31, 2002.

Lamarck, Jean-Baptiste. *Philosophie Zoologique*, 1809. Quoted in Stephen Jay Gould, *The Structure of Evolutionary Theory*. Belknap Press, 2002.

———. *Zoological Philosophy: An Exposition with Regard to the Natural History of Animals*, 1809, translated by Hugh Elliot. Macmillan, 1914. Reprinted by University of Chicago Press, 1984.

Lanois, Daniel. *Here Is What It Is*. Documentary film, 2007.

Lave, J. *Cognition in Practice: Mind, Mathematics, and Culture in Everyday Life*. Cambridge University Press, 1988.

Layden, Tim, and David Epstein. 'Why the Jamaicans Are Running Away with Sprint Golds in Beijing'. *Sports Illustrated* Web site, August 21, 2008.

Lee, Karen. 'An Overview of Absolute Pitch'. Published online at https://webspace.utexas.edu/kal463/www/abspitch.html, November 16, 2005.

Lehmann, A. C., and K. A. Ericsson. 'The Historical Development of Domains of Expertise: Performance Standards and Innovations in Music'. In *Genius and the Mind*, edited by A. Steptoe. Oxford University Press, 1998, pp. 67–94.

Lemann, Nicholas. *The Big Test: The Secret History of the American Meritocracy*. Farrar, Straus and Giroux, 1999.

Leslie, Mitchell. 'The Vexing Legacy of Lewis Terman'. *Stanford Magazine*, July–August 2000.

Levitin, Daniel J. *This Is Your Brain on Music: The Science of a Human Obsession*. Dutton, 2006.

Lewontin, Richard. 'The analysis of variance and the analysis of causes'. *American Journal of Human Genetics* 26 (1972): 400–411.

——. *Human Diversity*. Freeman Press, 1982.

Lieber, R. L. *Skeletal Muscle Structure and Function: Implications for Rehabilitation and Sports Medicine*. Williams & Wilkins, 1992.

Locurto, Charles. *Sense and Nonsense About IQ*. Praeger, 1991.

Longley, Rob. Column. *The Winnipeg Sun*, August 14, 2008.

Lowinsky, Edward E. 'Musical genius: evolution and origins of a concept'. *The Musical Quarterly* 50, no. 3 (1964): 321–40.

Ma, Marina. *My Son, Yo-Yo*. The Chinese University Press, 1996.

MacArthur, Daniel. 'The Gene for Jamaican Sprinting Success? No, Not Really'. Published on the Genetic Future Web site, August 21, 2008.

Maguire, Eleanor A., David G. Gadian, Ingrid S. Johnsrude, Catriona D. Good, John Ashburner, Richard S. J. Frackowiak, and Christopher D. Frith. 'Navigation-related structural change in the hippocampi of taxi drivers'. *Proceedings of the National Academy of Sciences* 97, no. 8 (April 11, 2000): 4398–403.

Makedon, Alexander. 'In Search of Excellence: Historical Roots of Greek Culture'. 1995. Published on the Chicago State University Web site, http://webs.csu.edu/~amakedon/articles/GreekCulture.html.

Malaspina, Dolores, et al. 'Paternal age and intelligence: implications for age-related genomic changes in male germ cells'. *Psychiatric Genetics* 15 (2005): 117–25.

Manners, John. 'Kenya's running tribe'. *The Sports Historian* 17, no. 2 (November 1997): 14–27.

McArdle, W. D., F. I. Katch, and V. L. Katch. *Exercise Physiology: Energy, Nutrition and Human Performance*. Williams & Wilkins, 1996.

McClearn, Gerald E. 'Genetics, Behavior and Aging'. Summary of BSR Exploratory Workshop, March 29, 2002.

——. 'Nature and nurture: interaction and coaction'. *American Journal of Medical Genetics* 124B (2004): 124–30.

McKusick, Victor A. 'Eye Color 1'. Published on the Online Mendelian Inheritance in Man Web site, National Center for Biotechnology Information, updated January 31, 2007.

Meaney, Michael J. 'Nature, nurture, and the disunity of knowledge'. *Annals of the New York Academy of Sciences* 935 (2001): 50–61.

Medvec, Victoria Husted, Scott F. Madey, and Thomas Gilovich. 'When less is more: counterfactual thinking and satisfaction among Olympic medalists'. *Journal of Personality and Social Psychology* 69, no. 4 (1995): 603–10.

Meltzoff, Andrew N. 'Theories of People and Things'. In *Theories of Infant Development*, edited by J. Gavin Bremner and Alan Slater. Blackwell Publishing, 2003.

'Men's Fidelity Controlled by "Cheating Genetics"'. *Drudge Report*, September 3, 2008.

Mighton, John. *The Myth of Ability: Nurturing Mathematical Talent in Every Child.* Walker, 2004.

Mischel, W., Y. Shoda, and M. L. Rodriguez. 'Delay of gratification in children'. *Science* 244 (1989): 933–38.

Mogilner A., J. A. I. Grossman, and V. Ribary. 'Somatosensory cortical plasticity in adult humans revealed by magnetoencephalography'. *Proceedings of the National Academy of Sciences* 90 (1993): 3593–97.

Montville, Leigh. *Ted Williams: The Biography of an American Hero*. Doubleday, 2004.

Moore, David S. *The Dependent Gene: The Fallacy of 'Nature vs. Nurture'*. Henry Holt, 2003.

———. 'Espousing interactions and fielding reactions: addressing laypeople's beliefs about genetic determinism'. *Philosophical Psychology* 21, no. 3 (2008): 331–48.

Morgan, Daniel K., and Emma Whitelaw. 'The case for transgenerational epigenetic inheritance in humans'. *Mammalian Genome* 19 (2008): 394–97.

Morgan, Hugh D., Heidi G. E. Sutherland, David I. K. Martin, and Emma Whitelaw. 'Epigenetic inheritance at the agouti locus in the mouse'. *Nature Genetics* 23 (1999): 314–18.

Morris, Edmund. *Beethoven: The Universal Composer*. HarperCollins, 2005.

Münte, Thomas F., Eckart Altenmüller, Lutz Jäncke, et al. 'The musician's brain as a model of neuroplasticity'. *Nature Reviews Neuroscience* 3 (June 2002): 473–78.

Murray, Charles. 'Intelligence in the Classroom: Half of All Children Are Below Average, and Teachers Can Do Only So Much for Them'. *Wall Street Journal*, January 16, 2007.

Murray, Charles, and Daniel Seligman. 'As the Bell Curves'. *The National Review*, December 8, 1997. Reprinted at http://eugenics.net/papers/mssel.html.

Murray, H. A. *Explorations in Personality*. Oxford University Press, 1938.

Myhrvold, Nathan. 'John von Neumann: Computing's Cold Warrior'. *Time*, March 29, 1999.

National Skeletal Muscle Research Center. 'Hypertrophy'. Published on the UCSD Muscle Physiology Laboratory Web site.

Neisser, Ulric. 'Rising Scores on Intelligence Tests: Test Scores Are Certainly Going Up All over the World, but Whether Intelligence Itself Has Risen Remains Controversial'. *American Scientist*, September/October 1997.

Neisser, Ulric, et al. 'Intelligence: knowns and unknowns'. *American Psychologist* 51, no. 2 (February 1996): 77–101.

Neumeyer, Peter F. *The Annotated Charlotte's Web*. HarperTrophy, 1997.

New Scientist Editorial Board. 'The Sky's the Limit'. *New Scientist*, September 16, 2006.

Nichols, R. 'Twin studies of ability, personality, and interests'. *Homo* 29 (1978): 158–73.

Nietzsche, Friedrich. 'Homer's Contest'. In *Five Prefaces on Five Unwritten Books* (posthumous writings), 1872. Available on The Nietzsche Channel Web site.

Nippert, Matt. 'Eureka!' *New Zealand Listener*, October 6–12, 2007.

Noakes, Timothy David. 'Improving Athletic Performance or Promoting Health Through Physical Activity'. World Congress on Medicine and Health, July 21–August 31, 2000.

November, Nancy. 'A French edition of Leopold Mozart's *Violinschule* (1756)'. *Deep South* 2, no. 3 (Spring 1996).

Nowlin, Bill. *The Kid: Ted Williams in San Diego*. Rounder Records, 2005.

Nowlin, Bill, and Jim Prime. *Ted Williams: The Pursuit of Perfection*. Sports Pub LLC, 1992.

Nunes, T. 'Street Intelligence'. In *Encyclopedia of Human Intelligence*, edited by R. J. Sternberg. Macmillan, 1994, pp. 1045–49.

O'Boyle, M. W., H. S. Gill, C. P. Benbow, and J. E. Alexander. 'Concurrent finger-tapping in mathematically gifted males: evidence for enhanced right hemisphere involvement during linguistic processing'. *Cortex* 30 (1994): 519–26.

Olympics Diary. 'Jamaicans Built to Beat the Rest'. *Dublin Herald*, August 19, 2008.

Oyama, Susan. *The Ontogeny of Information: Developmental Systems and Evolution*. Cambridge University Press, 1985.

Oyama, Susan, Paul E. Griffiths, and Russell D. Gray. *Cycles of Contingency: Developmental Systems and Evolution.* MIT Press, 2003.

Pacenza, Matt. 'Flawed from the Start: The History of the SAT'. Published on the New York University Journalism Web site.

Palaeologos, Cleanthis. 'Sport and the Games in Ancient Greek Society'. Published on the LA84 Foundation Web site.

Parable of the talents. Book of Matthew 25:14–30.

Pette, D., and G. Vrbova. 'Adaptation of mammalian skeletal muscle fibers to chronic electrical stimulation'. *Reviews of Physiology, Biochemistry and Pharmacology* 120 (1992): 115–202.

Phillips, Mitch. 'Jamaica Gold Rush Rolls On, US Woe in Sprint Relays'. Reuters, August 22, 2008.

Pigliucci, Massimo. *Phenotypic Plasticity: Beyond Nature and Nurture.* Johns Hopkins University Press, 2001.

———. 'Beyond nature and nurture'. *The Philosopher's Magazine* 19 (July 2002): 20–22.

Pinker, Steven. 'My Genome, My Self'. *New York Times Magazine*, January 7, 2009.

Plomin, R., and D. Daniels. 'Why are children in the same family so different from one another?' *Behavioral and Brain Sciences* 10 (1987): 1–60.

Pott, Jon. 'The Triumph of Genius: Celebrating Mozart'. *Books & Culture*, November–December 2006.

Powell, Diane. 'We Are All Savants'. *Shift: At the Frontiers of Consciousness* 9 (December 2005–February 2006): 14–17.

Pray, Leslie A. 'Epigenetics: genome, meet your environment'. *The Scientist* 18, no. 13 (2004): 14.

Quinn, Elizabeth. 'Fast and Slow Twitch Muscle Fibers: Does Muscle Type Determine Sports Ability?' Published on the About.com Sports Medicine Web site.

Raikes, Helen, et al. 'Mother-child bookreading in low-income families: correlates and outcomes during the first three years of life'. *Child Development* 77, no. 4 (July/August 2006): 924–53.

Ramachandran, V. S. 'Behavioral and magnetoencephalographic correlates of plasticity in the adult human brain'. *Proceedings of the National Academy of Sciences* 90 (1993): 10413–20.

Ramachandran, V. S., D. Rogers-Ramachandran, and M. Stewart. 'Perceptual correlates of massive cortical reorganization'. *Science* 258 (1992): 1159–60.

Rand, Ayn. 'Introducing Objectivism'. Times-Mirror, 1962.

Rastogi, Nina Shen. 'Jamaican Me Speedy: Why Are Jamaicans So Good at Sprinting?' *Slate*, August 18, 2008.

Reed, Edward S., and Blandine Bril. 'The Primacy of Action in Development'. In *Dexterity and Its Development*, edited by Mark L. Latash et al. Lawrence Erlbaum, 1996.

Rennie, Michael J. 'The 2004 G. L. Brown Prize Lecture'. *Experimental Physiology* 90 (2005): 427–36.

Ridley, Matt. *Nature via Nurture*. HarperCollins, 2003.

Risley, Todd R., and Betty Hart. *Meaningful Differences in the Everyday Experience of Young American Children*. Paul H. Brookes Publishing, 1995.

Rose, Tom. 'Can "old" players improve all that much?' Published on the Chessville.com Web site.

Russell, B., D. Motlagh, and W. W. Ashley. 'Form follows function: how muscle shape is regulated by work'. *Journal of Applied Physiology* 88, no. 3 (2000): 1127–32.

Rutter, M., B. Maughan, P. Mortimore, J. Ouston, and A. Smith. *Fifteen Thousand Hours*. Harvard University Press, 1979.

Rutter, Michael, Terrie E. Moffitt, and Avshalom Caspi. 'Gene-environment interplay and psychopathology: multiple varieties but real effects'. *Journal of Child Psychology and Psychiatry* 47, no. 3/4 (2006): 226–61.

Sacks, Oliver. 'The Mind's Eye'. *New Yorker*, July 28, 2003.

Sadie, Stanley. *The Grove Concise Dictionary of Music*. Macmillan, 1988.

Saretzky, Gary D. 'Carl Campbell Brigham, the Native Intelligence Hypothesis, and the Scholastic Aptitude Test'. Educational Testing Service Research Publications, December 1982.

Sarkar, S. 'Biological Information: A Skeptical Look at Some Central Dogmas of Molecular Biology'. In *The Philosophy and History of Molecular Biology: New Perspectives*, vol. 183. Kluwer Academic Publishers, 1996, pp. 187–232.

Schlaug G., L. Jancke, Y. Huang, et al. 'Asymmetry in musicians'. *Science* 267 (1995): 699–701.

Schönemann, Peter H. 'On models and muddles of heritability'. *Genetica* 99, no. 2/3 (March 1997): 97–108.

Shanks, D. R. 'Outstanding performers: created, not born? New results on nature vs. nurture'. *Science Spectra* 18 (1999): 28–34.

Shenk, David. *The Forgetting*. Doubleday, 2001.

———. *The Immortal Game*. Doubleday, 2006.

Sherman, Mandel, and Cora B. Key. 'The intelligence of isolated mountain children'. *Child Development* 3, no. 4 (December 1932): 279–90.

Shiner, Larry. *The Invention of Art*. University of Chicago Press, 2003.

Simonton, Dean Keith. *Origins of Genius: Darwinian Perspectives on Creativity*. Oxford University Press, 1999.

Slavin, R., N. Karweit, and N. Madden. *Effective Programs for Students at Risk*. Allyn and Bacon, 1989.

Snyder, A. W., and D. J. Mitchell. 'Is integer arithmetic fundamental to mental processing? The mind's secret arithmetic'. *Proceedings of the Royal Society of London. Series B, Containing Papers of a Biological Character* 266 (1999): 587–92.

Snyder, Allan W., Elaine Mulcahy, Janet L. Taylor, D. John Mitchell, Perminder Sachdev, and Simon C. Gandevia. 'Savant-like skills exposed in normal people by suppressing the left fronto-temporal lobe'. *Journal of Integrative Neuroscience* 2, no. 2 (2003): 149–58.

Society for Neuroscience. 'Mother's Experience Impacts Offspring's Memory'. Press release, February 3, 2009.

Spearman, C. 'General intelligence, objectively determined and measured'. *American Journal of Psychology* 15 (1904): 201–93.

———. *The Abilities of Man, Their Nature and Measurement*, 1927. Cited in Schönemann, Peter H. 'On models and muddles of heritability,' *Genetica* 99, no. 2/3 (March 1997): 97.

Spencer, J. P., M. S. Blumberg, R. McMurray, S. R. Robinson, L. K. Samuelson, and J. B. Tomblin. 'Short arms and talking eggs: why we should no longer abide the nativist-empiricist debate'. *Child Development Perspectives* 3, no. 2 (July 2009): 79–87.

Steckel, Richard. 'Height, Health, and Living Standards Conference Summary'. Published on the Princeton University Web site.

Sternberg, Robert J. 'Intelligence, Competence, and Expertise'. In *Handbook of Competence and Motivation*, edited by A. J. Elliot and C. S. Dweck. Guilford Publications, 2005.

Sternberg, Robert J., and Janet E. Davidson. *Conceptions of Giftedness*. 1st ed. Cambridge University Press, 1986.

———. *Conceptions of Giftedness*. 2nd ed. Cambridge University Press, 2005.

Sternberg, R. J., and D. K. Detterman, eds. *What Is Intelligence? Contemporary Viewpoints on Its Nature and Definition*. Ablex, 1986.

Sternberg, Robert J., Elena L. Grigorenko, and Donald A. Bundy. 'The predictive value of IQ'. *Merrill-Palmer Quarterly* 47, no. 1 (2001): 1–41.

Sterr, A., M. M. Muller, T. Elbert, et al. 'Changed perceptions in Braille readers'. *Nature* 391 (1998): 134–35.

Stowell, Robin. 'Leopold Mozart Revised: Articulation in Violin Playing During the Second Half of the Eighteenth Century'. In *Perspectives on Mozart*

Performance, edited by Peter Williams and R. Larry Todd. Cambridge University Press, 1991.

Strickland, Bonnie R. 'Misassumptions, misadventures, and the misuse of psychology'. *American Psychologist* 55, no. 3 (March 2000): 33–38.

Subotnik, R. 'A developmental view of giftedness: from being to doing'. *Roeper Review* 26 (2003): 14–15.

Suzuki, Shinichi. *Nurtured by Love*. Exposition Press, 1983.

Svendsen, Dagmund. 'Factors related to changes in IQ: a follow-up study of former slow learners'. *Journal of Child Psychology and Psychiatry* 24, no. 3 (1983): 405–13.

Symonds, John Addington. *The Life of Michelangelo Buonarroti*. BiblioBazaar, 2008.

Talent Education Research Institute. 'Personal History of Shinichi Suzuki'. Published on the Suzuki Method Web site.

Tauer, John M., and Judith M. Harackiewicz. 'Winning isn't everything: competition, achievement orientation, and intrinsic motivation'. *Journal of Experimental Social Psychology* 35 (1999): 209–38.

Terman, Lewis M. *The Intelligence of School Children: How Children Differ in Ability, the Use of Mental Tests in School Grading, and the Proper Education of Exceptional Children*. Houghton Mifflin, 1919.

———. *Genetic Studies of Genius: Vol. I, Mental and Physical Traits of a Thousand Gifted Children*. Stanford University Press, 1925.

———. *Genetic Studies of Genius: Vol. II, The Early Mental Traits of Three Hundred Geniuses*. Stanford University Press, 1926.

———. *Genetic Studies of Genius: Vol. III, The Promise of Youth, Follow-Up Studies of a Thousand Gifted Children*. Stanford University Press, 1930.

———. *Genetic Studies of Genius: Vol. IV, The Gifted Child Grows Up*. Stanford University Press, 1947.

———. 'The Discovery and Encouragement of Exceptional Talent'. Walter Van Dyke Bingham Lecture at the University of California, Berkeley, March 25, 1954.

———. *Genetic Studies of Genius: Vol. V, The Gifted Group at Mid-Life*. Stanford University Press, 1959.

Thayer, R., J. Collins, E. G. Noble, and A. W. Taylor. 'A decade of aerobic endurance training: histological evidence for fibre type transformation'. *Journal of Sports Medicine and Physical Fitness* 40, no. 4 (2000): 284–89.

Thistlethwaite, Susan Brooks. *Adam, Eve, and the Genome: The Human Genome Project and Theology*. Fortress Press, 2003.

Trappe, S., M. Harber, A. Creer, P. Gallagher, D. Slivka, K. Minchev, and D. Whitsett. 'Single muscle fiber adaptations with marathon training'. *Journal of Applied Physiology* 101 (2006): 721–27.

Treffert, Darold A. 'Is There a Little "Rain Man" in Each of Us?' Published on the Wisconsin Medical Society Web site.

———. 'Savant Syndrome: Frequently Asked Questions'. Published on the Wisconsin Medical Society Web site.

Treffert, Darold A., and Gregory L. Wallace. 'Islands of Genius'. *Scientific American Mind*, January 2004.

Trost, G. 'Prediction of Excellence in School, University and Work'. In *International Handbook of Research and Development of Giftedness and Talent*, edited by K.A. Heller, F.J. Monks, and A. H. Passow. Oxford: Pergamon Press, 1993, pp. 325–36.

Turkheimer, Eric. 'Three laws of behavior genetics and what they mean'. *Current Directions in Psychological Science* 9, no. 5 (October 2000): 160–64.

Turkheimer, Eric, Andreana Haley, Mary Waldron, Brian D'Onofrio, and Irving I. Gottesman. 'Socioeconomic status modifies heritability of IQ in young children'. *Psychological Science* 14, no. 6 (November 2003): 623–28.

University of the Arts. 'Paragone: Painting or Sculpture?' Published on the Universal Leonardo Web site.

Updike, John. 'Hub Fans Bid Kid Adieu'. *New Yorker*, October 22, 1960.

USA Today editors. 'In Every Sense, Williams Saw More than Most'. *USA Today*, June 6, 1996.

US News & World Report. 'World's Best Colleges and Universities'. Rankings based on the Times Higher Education–QS World University rankings, 2008. Published on the *US News & World Report* Web site.

Varon, E. J. 'Alfred Binet's concept of intelligence'. *Psychological Review* 43 (1936): 32–49.

Vasari, Giorgio. 'Life of Leonardo da Vinci'. In *Lives of the Most Eminent Painters, Sculptors, and Architects*, translated by Gaston du C. de Vere. Philip Lee Warner, 1912–1914.

Vineis, Paolo. 'Misuse of genetic data in environmental epidemiology'. *Annals of the New York Academy of Sciences* 1076 (2006): 163–67.

Von Károlyi, Catya, and Ellen Winner. 'Extreme Giftedness'. In *Conceptions of Giftedness*, 2nd ed., edited by Robert J. Sternberg and Janet E. Davidson. Cambridge University Press, 2005.

Wade, Nicholas. 'The Twists and Turns of History, and of DNA,' *New York Times*, March 12, 2006.

Wang, Yong-Xu, et al. 'Regulation of muscle fiber type and running endurance by PPAR'. Published on the Public Library of Science Web site, August 24, 2004.

Ward, P., N. J. Hodges, A. M. Williams, and J. L. Starkes. 'Deliberate Practice and Expert Performance: Defining the Path to Excellence'. In *Skill Acquisition in Sport: Research, Theory and Practice*, edited by A. M. Williams and N. J. Hodges. Routledge, 2004.

Wargo, Eric. 'The myth of prodigy and why it matters'. Association for Psychological Science's *Observer* 19, no. 8 (August 2006).

Watters, Ethan. 'DNA Is Not Destiny'. Published on the *Discover* Web site, November 22, 2006.

Weinberger, Norman M. 'Music and the Brain'. *Scientific American*, October 2004.

Weisberg, Robert W. 'Case Studies of Innovation: Ordinary Thinking, Extraordinary Outcomes'. In *The International Handbook on Innovation*, edited by Larisa V. Shavinina. Elsevier, 2003.

———. 'Expertise in Creative Thinking'. In *The Cambridge Handbook of Expertise and Expert Performance*, edited by K. Anders Ericsson et al. Cambridge University Press, 2006.

Whitwell, Giselle E. 'The Importance of Prenatal Sound and Music'. Published on the Life before Birth Web site.

Wierzbicki, James. 'The Beethoven Sketchbooks'. *St. Louis Post-Dispatch*, January 5, 1986. (Wierzbicki cites Johnson, Douglas, Alan Tyson, and Robert Winter, *The Beethoven Sketchbooks: History, Reconstruction, Inventory*. University of California Press, 1985).

Wilkins, John. 'Races, Geography, and Genetic Clusters'. Posted on the Evolving Thoughts blog, April 22, 2006.

Willoughby, Ian. 'Czech Ondrej Sosenka Sets New World One-hour Cycling Record of 49.7 km'. Radio Praha, July 20, 2005.

Winner, Ellen. 'The origins and ends of giftedness'. *American Psychologist* 55, no. 1 (January 2000): 159–60.

Winner, E., and M. Casey. 'Cognitive Profiles of Artists'. In *Emerging Visions: Contemporary Approaches to the Aesthetic Process*, edited by G. Cupchik and J. Laszlo. Cambridge University Press, 1993.

Winner, E., M. Casey, D. DaSilva, and R. Hayes. 'Spatial abilities and reading deficits in visual art students'. *Empirical Studies of the Arts* 9, no. 1 (1991): 51–63.

Wolff, Alexander. 'No Finish Line'. *Sports Illustrated*, November 5, 2007.

Wray, Herbert, Jeffrey Sheler, and Traci Watson. 'The World after Cloning'. *US News & World Report*, March 10, 1997.

Wright, Lawrence. *Twins: And What They Tell Us About Who We Are*. Wiley, 1999.

Wu, Echo H. 'Parental influence on children's talent development: a case study with three Chinese American families'. *Journal for the Education of the Gifted* 32, no. 1 (Fall 2008): 100–29.

Wyatt, W. J., A. Posey, W. Welker, and C. Seamonds. 'Natural levels of similarities between identical twins and between unrelated people'. *Skeptical Inquirer* 9, no. 1 (1984): 64.

Yang, T. T., C. C. Gallen, and B. Schwartz. 'Sensory maps in the human brain'. *Nature* 368 (1994): 592–93.

Yang T. T., C. C. Gallen, V. S. Ramachandran, et al. 'Noninvasive detection of cerebral plasticity in adult human somatosensory cortex'. *Neuroreport* 5 (1994): 701–4.

Young, Bob. 'The Taboo of Blacks in Sports'. *Willamette Week*, April 1, 2000.

Zaslaw, Neal, and William Cowdery. *The Compleat Mozart: A Guide to the Musical Works of Wolfgang Amadeus Mozart*. W. W. Norton & Company, 1990.

Zimmer, Carl. 'Now: The Rest of the Genome'. *New York Times*, November 11, 2008.

Acknowledgements

A person must first be alive in order to write a book, so I must first thank, quite literally from the bottom (and side) of my heart, Dr. Sidney Cohen, Dr. Robert Gelfand, and especially Dr. Manish Parikh. Humility is watching someone repair your own coronary artery live on a large television screen. I'm also grateful to Dr. James Blake, Dr. Phyllis Hyde, and Dr. Lawrence Gardner.

Somewhere in a first-year college writing class, a kid with more ability than I'll ever have is wondering if he could ever write books for a living. The answer is yes, if he never gives up and is lucky enough to get with the right people. I was lucky enough to get with Bill Thomas and Sloan Harris. This is the third book of mine that Bill has invested with excessive intellect and precious editorial resources. And, remarkably, it is the fifth book Sloan and I have conceived together (we used to talk to each other on dial phones). It is simply impossible to imagine my writing life without him.

I've already detailed, in the beginning of the Evidence section (page 143), how this book began and who helped inspire it in various ways. To those names, I must add profound thanks to Peter Freed, Patrick Bateson, and Massimo Pigliucci for early encouragement and vital insight.

As the work progressed, my draft readers kept me honest and clear. They are: Josh Banta, Patrick Bateson, Alexandra Beers, Mark Blumberg, Naomi Boak, Joanne Cohen, Sidney Cohen, Stan Cohen, Peter Freed, Rufus Griscom, Colin Harrison, Kurt Hirsch, John Holzman, Andy Hyman, Steven Johnson, Andrew Kimball, Gersh Kuntzman, Adam Mansky, Amani Martin, Massimo Pigliucci, David Plotz, Steve Silberman, Michael Strong, Francesca Thomas, Susie Weiner, and Sarah Williams. Jim Berman and Andy Walter took draft reading to a new level by reading and feeding me relentlessly.

For friendship and support, I am also indebted to Jeremy Benjamin, David Booth Beers, Peggy Beers, Eric Berlow, Carolyn Berman, Greg Berman, Chandler Burr, Bonni Cohen, Eamon Dolan, Bruce Feiler, Richard Gehr, Rob Guth, Andy Hoffman, Rachel Holzman, Steve Hubbell, Jane Jaffin, Roy Kreitner, Virginia McEnerney, Katherine Schulten, Andrew Shapiro, Jon Shenk, Josh Shenk, Richard Shenk(!), Leslie Sillcox, Mark Sillcox, Andras Szanto and Lea Thau. Special thanks to Anthony Uzzo and the lovely Hotel Beacon.

No agent or editor is an island. Profound thanks to Sloan Harris's outstanding crew at ICM: Kristyn Keene, Molly Rosenbaum, John DeLaney, and the great Liz Farrell; and to Bill Thomas's superb team at Doubleday: Maria Carella, Rachel Lapal, Sonia Nash, John Pitts, Nora Reichard, Alison Rich, and Amy Ryan. I am especially grateful for the patience and intelligence of Melissa Ann Danaczko.

Finally, the impossible task: to express in words the gratitude and pride I feel for my life-giving everythings, Alex, Lucy and Henry. Hopefully, they already know. The genius in all of us is that we each have the capacity to love and inspire one another.

Grateful acknowledgement is made to the following for permission to reprint previously published material:

Burkhard Bilger: Excerpts from 'The Height Gap: Why Europeans Are Getting Taller and Taller – and Americans Aren't' by Burkhard Bilger (*The New Yorker*, April 5, 2004). Reprinted by permission of Burkhard Bilger.

Malcolm Gladwell: Excerpt from 'Kenyan Runners' by Malcolm Gladwell (www.gladwell.com, November 16, 2007). Reprinted by permission of Malcolm Gladwell.

Jim Holt: Excerpts from 'Measure for Measure: The Strange Science of Francis Galton' by Jim Holt (*The New Yorker*, January 24–31, 2005). Reprinted by permission of Jim Holt.

Jennifer Keirn: 'Who's in Charge? Teach Kids Self-Control' by Jennifer Keirn (*Family Magazine*, July 2007). Reprinted by permission of Jennifer Keirn, www.jenniferkeirn.com.

Alexander Makedon: Excerpts from 'In Search of Excellence: Historical Roots of Greek Culture' by Alexander Makedon (Chicago State University Web site, http://webs.csu.edu/~amakedon/articles/GreekCulture.html, 1995). Reprinted by permission of Alexander Makedon.

Darold A. Treffert, MD: Excerpts from 'Savant Syndrome: Frequently Asked Questions' by Darold A. Treffert, MD (http://www.wisconsinmedical society.org/savant_syndrome/). Reprinted by permission of Darold A. Treffert, MD.

Giselle E. Whitwell. Excerpt from 'The Importance of Prenatal Sound and Music'. Published on the Life Before Birth Web site. Reprinted by permission of Giselle E. Whitwell.